With Data Driven Business Transformation, Carruthers and Jackson have managed to make data exciting for the second time! Let's face it, data has to be a precursor to any digital transformation. What resonates most, for me, is the ability to immediately start putting this into use -it's a how-to guide on roids!

The tangible model for assessing and transforming enables a blueprint for success. It's data for storytelling, not dictatorial; it's suggestive, open to a variety of different operating models, with hints of a Wardley-esque approach without being inflexible.

An easy read and digestible, written in language that is relatable without feeling like an old skool textbook. A testament to their love for the subject and how they bring it to life.

Rachel Murphy, Chief Executive Officer, https://difrent.co.uk

DATA-DRIVEN BUSINESS TRANSFORMATION

HOW TO DISRUPT, INNOVATE AND STAY AHEAD OF THE COMPETITION

CAROLINE CARRUTHERS
PETER JACKSON

WILEY

Registered office
John Wiley & Sons Ltd, The Atrium, Southern Gate, Chichester, West Sussex, PO19 8SQ, United Kingdom

For details of our global editorial offices, for customer services and for information about how to apply for permission to reuse the copyright material in this book please see our website at www .wiley.com.

Library of Congress Cataloging-in-Publication Data

Names: Carruthers, Caroline, author. | Jackson, Peter, 1962- author
Title: Data-driven business transformation : how to disrupt,
    innovate, and stay ahead of the competition / Caroline Carruthers, Peter Jackson.
Description: Chichester, West Sussex, United Kingdom : John Wiley & Sons,
    2019. | Includes index. |
Identifiers: LCCN 2018055952 (print) | LCCN 2018058802 (ebook) | ISBN
    9781119543220 (Adobe PDF) | ISBN 9781119543206 (ePub) | ISBN 9781119543152
    (hardcover)
Subjects: LCSH: Organizational change. | Management—Statistical methods. | Information
    technology—Management.
Classification: LCC HD58.8 (ebook) | LCC HD58.8 .J3345 2019 (print) | DDC 658.4/06—dc23
LC record available at https://lccn.loc.gov/2018055952

Cover Design: Wiley

Set in 10/14.5pt, PalatinoLTStd by SPi Global, Chennai, India.
Printed in Great Britain by TJ International Ltd, Padstow, Cornwall, UK

10 9 8 7 6 5 4 3 2 1

To Dez, Aidan and Jenny

CONTENTS

PREFACE

Bizarrely, because it has been around since what feels like the dawn of time, there are conflicting analogies about what data is and more critically how important it is to us. Data is often called the new oil; others dispute this and want to focus on the recyclable properties of data. Why is data even important when there are far more interesting things we could be talking about?

What data definitely is, is your base component. The building block on which the rest of your information and insight stands. Nothing, literally nothing, happens in any company without data being part of it. Some piece of data comes into play with every process, with every interaction. So why is the discipline of looking after and exploiting it such a relatively new concept?

We believe there are many reasons for this, including everyone believing that someone else is doing it—which never normally ends well! Businesses had acquired the habit of assuming that IT looked after the data, because it sits within the systems that they manage. However, IT thought that other areas of the organisation were taking care of the data, because that's where it was being generated and used. What was really happening was that a no man's land was being created where no part of the business was looking at the true value of data and what it could do for a company. Fundamentally, we don't see or treat data as a valuable asset to us. As people we literally throw data away, only using it in hindsight to indicate past performance through management reports, score cards and dashboards, for example.

Posts on social media now contain more detail than Stasi reports had during the Cold War. We tell shopping sites our purchasing aspirations in the hope that we might get a discount. As consumers we don't see any value in our data, yet to someone else it is extremely valuable, otherwise they wouldn't want it. Social media is a perfect example of this. We are not saying that social media is a bad thing, absolutely not. What we are saying is that we have noticed a trend where people are generally waking up to the value of what we all have and are starting to take this more seriously. This trend is also (thankfully) happening in organisations.

The advent of new technologies seems never ending. There is always something new and exciting to ooh and ah about. But—although we hate to be the voice of doom here—if we are creating all these new and wonderful things on a basis of poor data, then we are building a house of cards. They may stand and work perfectly, but we can trust them a great deal more if we have a solid data foundation to build them on.

We are always looking to the new, the bold, the visionary. It's exciting and who doesn't love a bit of adrenaline. That's natural—but let's try and set these things up to succeed by giving them legs.

Data is so fundamental to every company that we are definitely starting to see a shift in how it is viewed. The advent of the Chief Data Officer, which essentially elevates the importance of data in an organisation to that of an asset in the same way that money, people, structures and so on are, is proof that we are waking up from our data lethargy.

We need to discuss the difference between data-driven and data-enabled organisations. Data-driven organisations are either based on a data activity, such as Uber, Airbnb and Nectar; or wish to turn around their fortunes, driven by insights in data, to acquire more consumers, retain customers, increase customer lifetime value or decrease operational costs. This is transformation driven by the data. They are, in effect, data companies to some degree: data is their major asset.

Data enabled transformation is for organisations that wish to transform their business. By using data and data insights as the enabler for this (or one of the enablers), companies can better utilise their assets to support their wider business strategy. That said, the transformation they go through is data driven in order to become data enabled.

This book focuses on helping those companies become data enabled by showing the reader, step by step, how to take advantage of their data. Knowing that there is a problem but having no obvious way to fix it can be a scary situation to be in. This book is here to help with that providing a method for taking your company in the right direction to drive more value from assets you already have!

Whilst we give you an overarching method—so you know the big steps you should take and the direction to follow—because every company is different, we don't recommend you try and adhere to this prescriptively. If you can and do, then good luck and we hope it works for you. However, what might be more useful to you is taking the different strategies, tools and lessons from actually doing these transformations, and applying them in a way that resonates with your own company. There is no 'one size fits all'—and wouldn't life be boring if there was? Some areas will feel unwieldy for a fast-paced organisation, whereas others you may feel compelled to dwell on in order to end up in the right place for you. As long as you consider each area, so that any pace is a considered one, then you're good to get on with it.

One of the interesting things to take account of as you progress with business transformation is the transition from a programme mindset through to business as usual, and potentially back again. In order to begin change it is necessary to think in terms of what specific outputs you need to achieve (project thinking) in order to drive a certain outcome (programme thinking), so that the business can transform into a new 'business as usual'. This makes it sounds like a linear process, but unfortunately life is just not that straight forward. In reality you will end up flowing through different elements of this thinking as you progress, and once you think you are settled into a steady state then you will be looking at improving again.

Organisations can't afford to stay the same and change is part of business life. Different people will cope better with different parts of this thinking—and this is a really good example of why you need diversity of thought around you. If everyone thinks the same then they will be good at the same things and no one will be able to pick up the team when you reach an aspect that you aren't so great on. You don't fill a sports team with just one type of player, so why would you do this with your business teams?

There is a great deal of confusion around data terms. We do believe, however, that since everyone uses data on a daily basis this book should be easy to understand. To that end we have included a glossary of basic terms that are used throughout the book.

GLOSSARY OF TERMS

Analytics	The discovery, communication and interpretation of meaningful patterns in the data
Assurance	Activities to measure confidence in a given process, framework or data set
Audit	An exercise to determine compliance against governance documents and policies
Big data	Data sets that are so large and complex that traditional software can't deal with them efficiently. They have 'big' characteristics of three of the five Vs of data: volume, variety and velocity, (the other two being veracity and value)
Compliance	Adherence to requirements such as regulatory compliance
Data	A fact and a base component
Data architecture	One of the four enterprise architectures: a discipline focused on the models and policies that describe how data is structured, looked after and used
Data cleansing	The process of detecting and correcting corrupt or inaccurate records
Data dictionary	A catalogue and definition of all data elements
Data governance	The processes and framework that ensure important data assets are managed appropriately
Data lake	A single source/store of all the data within an organisation, often held as unstructured data

Data lineage	Describes where the data comes from, what happens to it and where it moves over time, often mapped between systems, applications or data stores
Data migration	The process of transferring data between storage types or systems
Data warehouse	A central repository of integrated data from disparate sources
Digital	The electronic technology that generates and processes data
Enterprise architecture	Made up of four architectures: application, business, data and system. This is a practice for analysing, designing, planning and implementing enterprise wide changes
Information	Derived from data that has been manipulated into something useful
Information architecture	A discipline focused on the design and organisation of information
Master data	A single source of common data used across multiple processes
Master data management	Curating and managing the master data to ensure its quality
Meta data	Data that provides information about other data: such as how long it is valid for, where, when and how it was created
Stakeholders	Someone (or a group) who is affected by a project or event

ACKNOWLEDGEMENTS

There are so many amazing 'data' people to thank that we could fill the book just doing that. For the sake of brevity special mentions have to go Emma Corbett, Alex Young and Tariq Bhatti who all contributed to our thinking in general and this book in particular. Adrian Wyman who worked his magic with his cartoons once again and the many, many wonderful people who we have both worked with and were happy to share their knowledge and experience with us for this book we hope you know how awesome you all are.

Thanks also has to go to Annie Knight and the amazing team at Wiley who have supported us from before day one right the way through the process. Your energy and professionalism have steered us through.

We would both also like to thank Aidan, Dez and Jenny who have given us the space and support to write this book. Finally, Caroline also has to thank her sisters and Dad who spend a lot of time telling her they have no idea what she is talking about but support her unconditionally anyway.

A very sincere thank you.

ABOUT THE AUTHORS

Caroline Carruthers is a recognised global data leader who is sought after for international speaking engagements. As well as co-authoring *The Chief Data Officer's Playbook* she is currently the Chief Executive of Carruthers and Jackson Ltd, the company behind data talks and data education initiatives such as 'the Chief Data Officers summer school'. She also helps clients to improve their data maturity and deliver data strategies within their organisations. Caroline has been a chief data officer across multiple industries and sectors and a data cheerleader since she realised how crucial it was to running a business well.

Peter Jackson is currently Director Group Data Sciences at Legal & General, having previously been Chief Data Officer at Southern Water and Head of Data at The Pensions Regulator. He is an accomplished speaker about all things data and the co-author of *The Chief Data Officer's Playbook*. Peter is passionate about data and the opportunities that data holds for the benefit of organisations and society.

Chapter 1

What is Transformation?

A great deal has been written and said about transformation, and many organisations are undertaking or experiencing transformation in one form or another. 'We have a number of transformation programmes' and 'we are delivering an organisation wide transformation over the next five years' are phrases often heard, and even better is 'we are always transforming'. Indeed, a whole industry has been built around scoping, designing and delivering transformation programmes.

'All organizations must undergo transformation to remain relevant. They must rethink customer experience, embrace digital business, and redefine business models and processes to remain competitive'.
Elise Olding Gartner Research VP and Conference Chair 2016

So, what is transformation? Transformation is a marked change in form, nature or appearance. Organisations that wish to transform are seeking to make a marked change in what they do or how they do it and the motive to do this is normally to remain relevant. Transformation is seen as being essential to remain competitive; gain market share; acquire, retain or reduce customer churn; or deliver services at a reduced cost.

More dramatic transformation can be, and has been, delivered by a range of different 'enablers' over time. Economic and social historians are able to catalogue a series of enablers: the harnessing of water power, the printing press, steam power and mechanisation, to name a few. Major transformations in society have been associated with the great 'revolutions' – for example, the British agricultural revolution of the 17th to 19th centuries and the industrial revolution of the mid-19th century. The most recent enabler of 'transformation' is what is known as 'digital' and this has been termed the third industrial revolution.

2

Obviously, these kinds of revolutions have a major impact on society as a whole and not just companies. But any company that doesn't embrace the new paradigm ceases to exist, which is a pretty good reason to embrace transformation.

For an individual organisation to undergo transformation of any kind there are risks and costs that are perceived to be outweighed by the benefits of the end state after the transformation. There are some very well documented examples of organisations that have failed to take the opportunity to 'transform'.

'Kodak did not fail because it missed the digital age.
It actually invented the first digital camera in 1975.
However, instead of marketing the new technology, the
company held back for fear of hurting its lucrative film
business, even after digital products were reshaping
the market'.
Avi Dan (January 2012) CMO Network

Kodak, for example, feared the risks of hurting its film business and failed to transform. Rather than focus on the possibilities of the digital age and embrace a new market it held fast to the past and completely missed its opportunity, becoming irrelevant.

Transformation is risky, painful, costly, requires change and needs great leadership. Therefore, transformation is often avoided in favour of smaller less useful exercises or does not deliver because the key elements aren't in place.

It is often easy for organisations to talk convincingly and even passionately about transformation without embracing it or delivering the transformation end state. There are many reasons why transformations fail and these will be discussed later in the book. Do organisations really 'transform'? Is the end state so different from the start position that the change can be described as 'transformative'? We would suggest that in the majority of cases the answer is probably no. For companies to undergo transformation they must 'disrupt' the way they do things. The way they operate, their operating model, the products and services they offer, their relationship with their customers or regulators, or perhaps their relationship with their own employees, in most cases they want to protect these things rather than disrupt them.

To achieve transformation the new end state must be imagined. The vision must be more than built, described and understood, it must be brought to life in a compelling fashion that makes organisations want to go through the pain because it's worth it. They must be prepared to hold their course when the going gets tough. Lots of work and effort need to be put into understanding the new desired end state, to understand its feasibility or its shape.

During any kind of transformation data is a useful tool to help achieve the transformation; however, it may not be the central focus for the trans-formation. In simple terms, data is used to build the business case. So, from the very inception of the change data is used as an enabler. In fact, the insights in the data may actually reveal that change is required and be the transformation driver.

Take the example of a business reviewing its monthly sales, by looking in the rear-view mirror, using management reports. It may see that its monthly sales figures are down. Perhaps the projections for next month's sales are more positive and suggest that sales may pick up. How many organisations though will project out the sales figures, revenue and costs over an extended period, say five or ten years? The business may be making a profit even on declining sales, they may be propping up this profit by cutting costs (and perhaps quality), but how many companies would look far enough into the future to see when the sales hit the x-axis? If an organisation uses the data to look that far ahead, they may realise that the data is showing them that they need to transform their operation.

Another example may be a water utility company in an arid region faced with growing demand for water and reduced supply from climate change. Projected out of long time scales, taking in many factors such as changing land use and standard of living, the data may provide insights showing that they need to transform their business to manage the supply–demand relationship in the future.

Once the new end state has been developed into a vision and a strategy, data continues to play a part throughout the transformation process. Key performance indicators (KPIs) will be created to measure the progress of the transformation. Data will drive these KPIs and be used to drive the decisions of the executive and transformation delivery team throughout the process.

Insights provided by the data are embedded into the decisions that drive or enable the new end state, whether these are customer, product, supply chain or operational insights.

But what really interests us in this book are the situations when data is at the heart of transformation end state. We will go on to talk about this in a bit more detail.

Why data and not digital?

Some organisations speak of 'digital transformation', in the main because we have been confused about what digital actually is. It is a frequently used term that is commonly mis-understood. We recently attended a conference in London focused on transformation. A panel discussion in the programme was titled:

'What impact will digitisation have on driving customer acquisition and retention?'

However, the slide behind the panellists on stage had the title:

'What impact will digitalisation have on driving customer acquisition and retention?'

It's a one-word difference between the two questions, but it had the potential to transform the discussion. The starting question should have been what is the difference between 'digitisation' and 'digitalisation'? However, this was not addressed at all, in fact it was ignored.

So, let's explore these terms, because understanding them will help us understand the role and importance of data in driving these outcomes and therefore driving transformation.

There are three similar terms involved in 'digital' and it helps to understand these to unravel different types of enablers behind transformations. Often an organisation's desired transformation fails to deliver because there are misconceptions about the type of transformation that is being undertaken and what the true enabler is.

Digitisation

The first is digitisation. This is the conversion of analogue information into digital form. An example of this is the automation of existing manual- and paper-based processes. Another example is the conversion of a physical photograph into a digital image either by scanning or re-photographing the photograph using a digital camera. So, digitisation is perhaps the starting point on the 'digital' journey. There isn't any real change going on but rather a repeat of a physical process, action or picture in the digital world. No change actually takes place on the artefact in question.

Digitalisation

The second is digitalisation. Unlike digitisation, digitalisation is the actual 'process' of the technologically induced change. It is the use of digital technologies to change a business model. This is what most people mean when they say 'digital' transformation. The use of new platforms to change operating models and drive new revenue or operational efficiencies. In this process you are still really focusing on the tools rather than taking a more holistic approach: it's like buying a car and not worrying about having a driving licence, insurance or wheels.

Digital

Finally, digital transformation is described as the total and overall societal effect of digitalization.

> *Digitization has enabled the process of digitalization,*
> *which resulted in stronger opportunities to transform*
> *and change existing business models.*
> *Digitization (the conversion), digitalization (the*
> *process) and the digital transformation (the effect)*
> *therefore accelerate and illuminate the already existing*
> *and ongoing horizontal and global processes of*
> *change in society.*
> Khan, Shahyan (16 September 2016) *Leadership in*
> *the Digital Age – a study on the effects of digitalization*
> *on top management leadership – quoted in*
> *Wikipedia*

So, returning to the conference we attended, was the question about the effects on customer acquisition and retention caused by moving from paper-based processes to digital processes, that was something that would have been discussed by a panel in 1990? Or was the question about the effects on customer acquisition and retention generated by use of digital platforms? Perhaps the question being put to the panellists was really about the effects on customer acquisition and retention caused by the use of data in a digital ecosystem.

We know what we think, and it was interesting that the panellists answered this from a technology point of view, rather than from a data-driven insight one. They discussed online experience and interaction and didn't talk about insights driven by data and personalisation created by data insights. They were all missing parts of the puzzle.

> *'Two-thirds of all business leaders believe that their companies must pick up the pace of digitalization to remain competitive.'*
> Susan Moore (30 October 2017) Embrace
> the Urgency of Digital Transformation, Gartner

At the Gartner Symposium/ITxpo 2017 on the Gold Coast, Australia, Val Sribar, Senior Vice President at Gartner said that

> *'Many businesses are stuck running digital projects. Some of them are very large, but digital projects are not a digital business. Executive leaders are frustrated with the slow pace of digital transformation, as they watch competitors capture new opportunities. CEOs are looking to CIOs to create new efficiencies, new value and new ways to engage constituents using*

*technology, quickly. Two-thirds of all business leaders
believe that their companies must pick up the pace of
digitalization to remain competitive. However, digital
transformation is nearing a period of intense scrutiny.
Four years into the digital shift, we find ourselves at
the "peak of inflated expectations", and if the Gartner
Hype Cycle teaches us anything, a trough is coming.
Disillusionment always follows a period of
extreme hype'.*

The question must be asked; why isn't digital delivering the business outcomes, the transformation outcomes, that are expected? The answer is because the focus is on a 'technology digital transformation' (the platforms) rather than a 'data digital transformation' (the insight from the data). Technology in itself does not deliver value, it is how that technology is used. To coin a very old phrase, it's not what you have it's how you use it that matters.

Michele Caminos, managing vice president at Gartner, said 'the key to speeding through the trough of disillusionment and creating value at scale is all about people'.

Gartner has identified a looming talent gap for key technology skills in artificial intelligence (AI), digital security and the Internet of Things.

*'But it's not just IT jobs', Caminos said. 'There's been
a 60% growth in technology skills required for non-IT
roles over the past four years. Digital business also
requires a new set of attributes and skills that allow
you to operate successfully in a continuously changing
world, like more frequent complex decision making,*

9

continuous problem solving, rapid pattern recognition and exception handling.'

Susan Moore (30 October 2017) 'Embracing the Urgency of Digital Transformation', *Smarter with Gartner*

These attributes are all about 'data' and not 'digital'. Digital transformation has been focussed on equipping an organisation with technology, digital platforms; or 'tooling up an organisation'. The data has often been overlooked. A new piece of technology in itself will not deliver transformation. The solution to a problem is not the tool (or the technology) it is how that tool is used. In other words, it is often the data being collected, stored and processed in a digital technology that delivers the insight and information to transform a business. It is the data that provides the actionable insights or the predictive analytics, it is not the technology.

Capital One has invested in developing and upskilling staff. Its talent programme focuses on key technology disciplines. In July 2018 M&S set up their data academy to train more than 1,000 staff in 'data skills', that is, machine learning and AI. This was billed as

'the biggest digital investment in staff to date as part of the ongoing efforts to ensure the business is fit for the future.'

M&S is ramping up its efforts to become a digital-first business with the launch of a new skills initiative that is looking to create "the most data-literate leadership team in retail".'

Ellen Hammett (30 July 2018) *Marketing Weekly*

M&S probably had fit for purpose digital platforms but was missing the 'data piece'. It is interesting that M&S boss Steve Rowe said 'We need to change our digital behaviours, mindsets and our culture to make the business fit for the digital age'. In order to do this what he actually trained his staff in was data literacy.

We can already see that the fourth industrial revolution will be data driven. To recap on the revolutions: the first industrial revolution took place in the 18th and 19th centuries and involved innovations such as steam engines and mechanical production. The second industrial revolution was the end of the 19th century and just before World War I. It included advancements such as the telegraph and industrial sewing machines. The late 1970s saw the third industrial revolution, which is still ongoing and has been powered by the internet and smartphones.

The fourth industrial revolution will be a future of AI and robotics, and its lifeblood will be data, perhaps big data, but big data is just data on a large scale. According to Robert Dagge, big data is 'Industry 4.0' (4 November 2016, Global Manufacturing).

The Fourth Industrial Revolution Report delivered at Big Data London 2017 concluded that it is clear that the need to become data driven is a global imperative. Companies have to transform their cultures and data architectures to adapt to new technologies, new types of data and, importantly, to new governance and compliance regulations.

Before moving on in the discussion, the shift away from a technology driven digital transformation towards a data enabled transformation is emphasised by Klaus Schwab, chairperson of the World Economic Forum:

'In the future, talent, more than capital, will represent the critical factor of production'.

In the future it will be less about the capital to buy digital technology and more about the talent and skills to harness the power in the data. Data is so much more than the new oil, it is the sustainable endless energy source we have been looking for. But if this is the case then how can the energy in this powerful asset be unleashed?

The focus for successful transformations should be the data and not the technology. Let's start using the correct words and the outcomes may be more successful: you have a data driven transformation to create a 'data enabled organisation'.

Drivers

Data is literally everywhere now, it has such an impressive impact on our lives, improving the quality from birth right through to technology using data that improves quality of life towards the end. Using massive data sets has revolutionised everything from how we watch TV, to going on holiday, to traveling in taxis, but it has had more fundamental impacts as well. Tracking weather data, for example, means we can predict extreme weather events, prepare for them and potentially save lives. Just imagine the data sets that mean the elderly can stay in their homes for longer, giving them a better quality of life

That said, you don't need be trying to save the world to get better use of your data. If you are stuck in a hell of endless spreadsheets that never match it can feel pretty life changing when it eventually gets worked out. However, you have a few things to sort out first. There a number of hurdles an organisation faces when it comes to data:

1. You have multiple data sources but no control over them. You aren't even sure where some of the data is coming from or why you are using it. Different departments source data from external agencies without looking at the overall impact on or needs of the business.

2. You have different versions of data, so you think you are looking at the same thing but you are actually looking at competing data and no actual master data exists.

3. Your tools seem to work against you not for you. Poor infrastructure means you spend more time trying to fix your data or work around it rather than it enabling you.

4. There is a lack of trust in the data, which could be real or perceived (in most cases it's a blend of both).

5. You don't have clear roles and responsibilities around data.

6. There are problems with the scale of data, you literally have too much data that most people in your organisation don't understand the value of.

7. You have integration complexity so nothing matches or you have brought companies together but don't truly understand the complexity of integrating the data so you never get over the first barriers.

8. A lack of governance means that even trusted data can be manipulated without control and lose its integrity.

9. Unpredictability of data and availability: if you can't rely on it you find different ways of performing your role, which can take a lot longer.

10. There is a need to work at different levels with the data, down in the detail and up in the stratosphere. However, each level is building what it needs from scratch rather than looking at the same data through a different lens.

The drivers for data-enabled transformation for organisations have been covered to some extent already. However, it is worth drawing these out. Regular engagement with our fellow chief data officers (CDOs) and chief information officers (CIOs) from numerous organisations shows that there are some common drivers pushing or drawing organisations into data-driven transformation. These drivers tend to come from one or more of the following five origins.

The first is competitive advantage. Many organisations understand that their data contains insights that will deliver a competitive advantage. An advantage that will allow them to serve their customers better and more efficiently. The advantage may be to increase customer acquisition, reduce customer turnover, increase customer lifecycle value or match products and services to customers at the right time in the right place. This driver includes not only those organisations seeking to transform to grow, but also those organisations looking to transform to survive. Think about the basic TV channels and what happened when Netflix came along.

The second is customer service and personalisation. Again, organisations may appreciate that the data they collect, or could collect, provides the raw material to give a better more personalised customer experience This experience may be delivered on a digital platform or in an analogue manner, again exposing that the transformation is less digitally enabled and more data enabled. This model is one that the personal insurance market works with really well.

Third is operational efficiency. Most organisations can improve operational efficiency through observations in their data. Data analysis or advanced analytics can provide actionable insights for informed decision making that could transform an organisation. Large transportation companies and utilities are starting to make the most of this area.

Fourth is regulatory pressures and requirements. Organisations, and this is the majority now, operate in a regulated environment, where they are required to provide regulatory data for compliance or assurance. Transformed operating models based on data and data insights can make the regulatory environment more efficient, more assured and change the relationship with the regulators.

Perhaps the final driver arises because other forms of transformation have been attempted and not delivered the business outcomes that were required. This is far from unusual. A remarkable number of organisations in all sectors and verticals have undergone, often very costly, technology or digital transformations that have failed to deliver tangible business outcomes. In some extreme cases the 'transformation' has delivered a negative outcome. This may be as a result of starving other initiatives of capital investment, causing significant business disruption during the 'transformation' or simply by making the technology more complicated and less resilient than before.

If these are the basic economic or commercial drivers, others may be more abstract in nature. The driver may originate from a CEO, or senior executive, understanding the true potential that lies in the organisation's data. This visionary may be in a position to champion and instigate a data-driven transformation. This person may couch it in terms of preparing for the future or getting ahead of the competition, or even disrupting the business from within rather than being disrupted by external forces. However, this usually happens because that person is addressing one of the five above-mentioned areas.

Whatever the driver, organisations are beginning to realise that transformation is driven by more than technology and platforms. In some ways the data is the new oil analogy does work. The combustion engine is the platform, the technology, the petrol is the data. Put the data into technology and deliver the right spark (the people) and energy is released. The engine alone will not produce the energy. On its own the petrol can produce energy with the right spark, but this energy will be released in an uncontrolled and chaotic manner, an explosion. Put the data (the petrol) and the technology (the engine) together and controlled powerful energy is released. Just remember that what you have with data is a much more powerful asset than oil, which you can reuse and reuse.

Data underpins so much of your organisation that to transform without paying attention to it means that you are building a house without clear foundations, it might work but do you really want to take that chance?

Who should drive?

Most business transformation for the past two decades has been driven by technology, and therefore by the chief technology officer (CTO) or CIO. Much transformation has been 'done to' the business so that staff don't feel part of it: and human beings are really good at digging their heels in when they feel like they don't have a say in what is happening to them.

It is interesting to note that often the project or programme delivery area sits within the information technology (IT) function. Technology is brought into the business to deliver transformation. Has this worked? If it has, why do so many technology-led transformations fail to meet the expectations of the business, or just fail? Also, if the past two decades of technology-driven transformation have been successful, why are so many businesses currently undertaking transformation programmes?

If we look at our experience in the 1980s (and we are not talking about the big hair and rather questionable dress sense, but IT and what it was like to be part of it), IT was literally the BIG thing and, as a result, lots was made of IT-driven organisations. If you didn't have an IT department full of enterprising people working on projects then your company was obviously going nowhere. Quite a few of the support people existed in basements answering questions on why people couldn't add another column on to a behemoth of a spreadsheet in scenarios that were just a bit too close to *The IT Crowd* for comfort.

IT departments trebled in size and almost became the reason companies existed – almost. What really happened is that the IT staff became comfortable in their role as company rock stars and began to forget what the purpose of the organisation was. Software and hardware were exciting and we started inventing reasons to buy things rather than focusing on what we were trying to achieve. Genuinely we tried to make sense of this situation and put in place roles like business analysts to form a translation service between the business side of the company and the technical services, while IT departments spawned more and more differing job titles to cope with evolving disciplines.

Companies became IT driven, forgetting what they were supposed to be focusing on and instead working on new shiny things. When as an industry we realised what was happening we tried harder to work with the business to get the best from the different parts (it really brings meaning to the sum of the parts being better than the individual parts). Since then the emphasis has been on creating IT-enabled companies, where IT creates a platform for the rest of the company to work on, enabling them to be better.

Despite the fact that looking after data properly is a relatively new discipline, the number of CDOs is steadily increasing and people are moving into data and information related roles while we pioneer and formalise what we do. If we aren't careful, we will fall into the trap that IT fell into in the 1980s – where we start to focus on all the things that we can do with data rather than the data strategy feeding the business strategy. Placing all of your focus on the data is fine if you are a data company, not so much if you are making socks.

We are really clear that the data professionals in a company are not the rock stars, rather they create the environment for others to be the rock stars. So, what we want is to turn a company into a data-enabled company where the focus is on the business strategy and how the data strategy underpins and helps to deliver it. It is the transformation that is data driven, the focus of the transformation is the data and how a company looks after it so it is right that the transformation is data driven.

What about the transformation itself? For something to be transformation it has to have a marked change in form, nature or appearance – and the change word is very scary for companies. Something that creates a marked change is radical, which equates to mega frightening, but people are excited by the idea of using data (which they already have an abundance of – possibly a byproduct of hoarding too much but we deal with that in another chapter).

Who should drive the data-enabled transformations? The straightforward answer is the business. In simple terms the business units, whether that be operations, manufacturing, service delivery, engineering, investment, financial services, human resources (HR), procurement or finance. It is the business that should drive the transformation. In many instances the organisation needs more data, more insight to power, in order to enable or sustain the transformation. The company will often need data to answer the questions that they are facing. At the end of the day, the point of the change is to make the business work better, more efficient, more effective and so on, not to put in place a new shiny tool. Therefore, shouldn't the people who are committed and focused on that goal not be the ones to drive the change?

A question that often gets asked at conferences and round-table events is how data insights can be embedded or operationalised into the organisation. The answer we always give to this question is that the business should be involved from the very start. Work the company throughout the initiative or project so that the data insights become native to the organisation and are adopted from the very start of the project. To truly enable this, an agile approach to delivery lends itself to data-driven transformations, working closely with the business through a series of iterations to reach the end state. Data and data technologies tend to lend themselves to this style of working, moving towards the longer-term big solutions of the business as usual elements increase. The more transactional something is, the less focus there should be on agile as you improve it. Not everything has to be completed in the same fashion.

The approach of the business driving the data-driven transformation is very different from the old school technology-driven project management where the organisation would provide the 'sponsor' of the project. This in itself gives it all away. The business is sponsoring a technology transformation but the sponsor rarely has a real hands-on role with the transformation – usually this is monthly check-ins and a lot can happen in a month. This should be reversed, with the business playing the major part in guiding the change. In a data-driven transformation, the data is supporting a business transformation to enable the company to get more from their data assets. You could almost see the CDO as the sponsor, forming the data and data insights into the business transformation.

The support functions must let the business lead and most importantly listen to them. Data enabling an organisation involves the data leaders listening to the needs and the questions that the business is asking.

The traditional triangle of people, processes and technology (see Figure 1.1) has been disrupted by data.

In our previous book *The Chief Data Officer's Playbook* we proposed a new relationship, as shown in Figure 1.2.

We now understand that data enabled businesses truly disrupts this model and the model of business transformation so that, as in Figure 1.3, data is the driver behind the transformation.

The relationship between the elements of transformation change once the business is the driver and the transformation is data enabled. As discussed previously, the business uses data and insights from data to help shape and make decisions about the future end state vision. This will

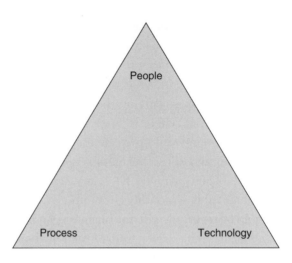

FIGURE 1.1 Business ecosystem triangle

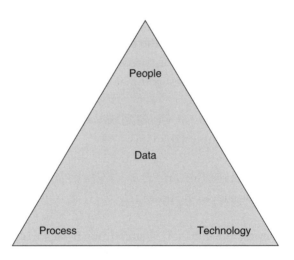

FIGURE 1.2 Business ecosystem triangle incorporating data

require much iterative working, testing the insights, tuning them and the models. Once a future end state vision has been developed the technology requirements, people requirements and process changes are defined and the business transformation is delivered to reach that end state vision.

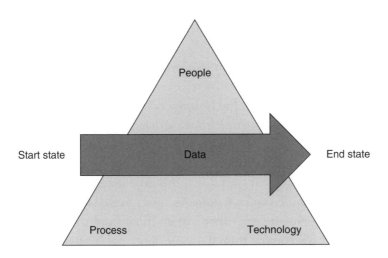

FIGURE 1.3 Business and data ecosystem transformation

At this point the data sustains the end state through data collection, data processing and data analytics. There will be a feedback loop of data that might run the transformation process again, which is possible if an agile approach has been adopted or will be constantly finetuned, providing continuous improvement to the transformed business. We will talk about this in much more detail in Chapter 2.

There may be some debate about whether the transformation function should sit independently outside of the business, outside of technology and outside of data. In an independent role the transformation may be able to pull together and coordinate multiple programmes that are elements of an overall transformation programme. This may be equally valid, especially in some larger and more complex organisations. However, the principle suggested at the start of the chapter remains constant. Transformation should not be driven by technology. Going back to first principles the technology itself tells us nothing: it does not provide insights or tell stories (we will come onto this in Chapter 2). It is the data in or being delivered from the technology that provides the insights and enables us to tell the stories that will transform businesses.

Data literacy – the art of data

Data literacy is vitally important to power data-driven business transformation. Data literacy is the ability to understand and communicate data as information, focusing on the competencies involved in working with the data. It has a wide spectrum, from the data scientist who can take data sets and create insight or information to the data citizen who can understand, appreciate and act on those insights because they understand where the insights come from. It's a bit like different dialects within the same language: two people can be talking the exact same language but find it difficult to understand each other if they come from different places. It helps that they are coming from the same base, that is, speaking the same language, but they should be careful of the nuances.

We think we can all accept that society today needs to be more 'data literate' to understand how their data is being used and abused. Data often has a value transaction associated with it. The data-literate citizen will understand this value. An individual provides their data to their social media platform and in return receives the utility of an enterprise grade communication and media platform: but how many people really understand what they are signing up for? In an increasingly data-driven world, one that is continually exploited by data, it becomes increasingly important that everyone – citizens, businesses and governments – becomes data literate. The shift to data-driven transformation has happened so quickly that huge parts of society have been left behind in their comprehension of the power of data. Since 2010 there is evidence that democracy has been increasingly disrupted by the use of data. Perhaps democracy is undergoing data-driven transformation? In this world everyone needs to be data literate.

It is essential for the data-enabled business, and therefore for data-driven business transformation, to have a high level of data literacy across the organisation. We can refer back to the initiative taken by M&S in July 2018 to start the 'data skills' training of more than 1,000 staff to establish 'the most data-literate leadership team in retail'.

> *'Putting data in the hands of a few experts is a powerful thing, but making it available throughout the organisation can be a game-changer'.*
> Doug Bordonaro (1 March 2018) InfoWorld

Creating a data-literate organisation will enable data to be put into the hands of the whole organisation. Data literacy will allow the democratisation of data across an organisation, which can then power transformation. In simple terms think about the power of every single member of a company all pushing the company in exactly the same fashion – don't underestimate the power of the crowd. As you can see from the data-literacy spectrum in Figure 1.4 there is no one type of person who will make your organisation data enabled. Rather, as well as your data specialist, you will need to harness a range of talent from you data citizens to your data-aware people who have the lightest touch points with your data.

An organisation that does not have a data-literate work force will meet challenges going forward, and will struggle with transformation. The data literacy has to be throughout the organisation from the highest levels, for

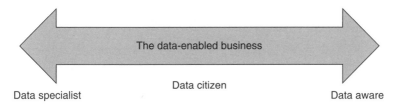

FIGURE 1.4 The data literacy spectrum

example the board and the executive team through management to the shop floor and operational levels. No one is exempt from improving their data literacy.

We often refer to the DIKW pyramid: data, information, knowledge and wisdom (see Figure 1.5).

To help understand the context of data literacy and where this sits within an organisation we have refined the DIKW pyramid to become the data-enabling pyramid, which we think makes it clearer from an organisation's point of view (see Figure 1.6).

Data literacy sits across the top end of the data layer through what was the information layer and into the lower parts of the knowledge layer. We have changed the name of the information partly to disentangle this from IT, where it causes confusions with 'information security and information management'. This layer is better described by the three 'Cs': data that has been collated, curated and contextualised to provide insight or information for the business. The three Cs (or as we lovingly call it C^3) are

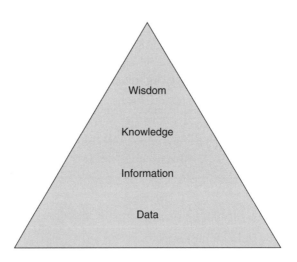

FIGURE 1.5 DIKW pyramid

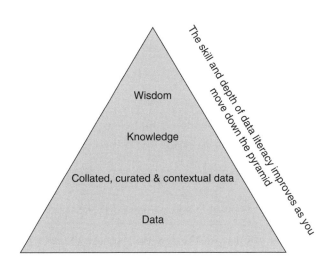

FIGURE 1.6 Data-enabling pyramid

performed by the 'data professionals'; the data governance department, engineers, architects and scientists. It helps to understand the layers in a bit more detail.

Layer 1: data

This is raw data, unprocessed data perhaps in a very granular form, with a lot of 'noise'. This is data that is sitting very close or in the operational technology (OT) function. It may have been extracted but even so it probably remains in a very raw, or even native, form. To gain any insight from this type of data is a very manual process and likely requires a great deal of domain knowledge and experience. It literally will be found by looking at the data, searching through the numbers – needle in a haystack approach.

Layer 2: C³

In this layer the data is managed – collated, curated and contextualised – not by data architects but by 'data wranglers' and is brought into a form that makes it of use to the business and data scientists. In this layer the data is processed to be fit for the purpose of enabling knowledge.

Layer 3: knowledge

The knowledge layer is where subject matter experts, those with domain knowledge, use the C^3 data to create the knowledge and the insight to drive the business.

Layer 4: wisdom

This is the furthest point from the raw data where the insights are being used to create a body of wisdom to run the business.

The need for data literacy sits across all four layers, but the depth and skill in data literacy decreases as one progresses up the pyramid. Often an organisation needs the CDO to tell the story here because they have the skills and knowledge across the whole pyramid.

Alongside all of this your data team provides the agility and interaction to help the flow across the layers – remember this is not a linear process, you move fluidly and by jumping around the pyramid.

To put it in really simple terms:

Data is knowing there is a red light.

Information is knowing that the red light sits above amber and green lights, which are not on.

Knowledge is knowing that that configuration is a traffic light.

Wisdom is knowing that you are heading for it and that you have to stop.

Data literacy

What is data literacy? Because of the wide range that it needs to span across the pyramid it might be hard to define in an all-encompassing sentence. At the data professional end of the spectrum data literacy is more about data skills. These skills would include data modelling, data engineering, the ability to extract, transform and load data, to manage and govern data. These are skills involved with master data management and meta data; skills required to build data catalogues, data dictionaries and data lineages; the ability and skills to create data visualisations, automated dashboards and reports; data scientist code skills of R and Python, statistical skills and building algorithms. These are true data professional skills that would not be expected at the business end of the data literacy spectrum. At the other end of the spectrum the data aware would be first to understand that these data professional skills and processes exist and how they are important. Then more specifically to be able to understand the data and collated information and draw insight from that rather than drawing exclusively upon 'gut feeling' or 'experience' to make decisions.

Whatever the level of skill required to be data literate, an important element of data literacy is the ability to tell a story from the data. This is the most powerful business skill. Technology platforms do not provide any insight, equally raw data or processed data do not tell a story. The story is drawn out by a subject matter expert who is data literate. Even machine learning and AI require a subject matter expert who is data literate, or a data scientist who is business literate, to draw out and tell the story.

Whatever sits in the detail of data literacy at any point in the pyramid, the most important feature of data literacy that must be adopted by any organisation wishing to adopt a data enabled transformation is a 'data culture'. An organisation has to build and adopt a data culture. This means valuing the data within the organisation, managing, owning, sharing and seeking to exploit the data. A well-developed data culture will power a data-driven transformation.

> *'Achieving data literacy has several components. Tools and technology are part of it, but employees must also learn to think critically about data, so they can understand when it's useful and when it's not. And perhaps most importantly, data literacy requires a culture in which data is valued by all as a primary vehicle for decision making.'*
> Doug Bordonaro (1 March 2018) InfoWorld

Bordonaro goes on to suggest that there are four enablers for data literacy:

1. Widespread and easy access to data. Importantly this must be access to data that is governed and trusted.

2. Leadership onboarding a data culture.

3. A platform for sharing data, most importantly trusted, governed and consistent data.

4. Critical thinking.

Because data literacy has such a huge span from very specific data skills, to culture and storytelling, the term data literacy might not be the correct term. The range of skills and knowledge required may be better described as the 'art of data'. The best artists have the technical skills to create artworks, but the citizen is equipped with enough understanding to appreciate and value the art (though they may not necessarily like the art) and can tell the story in it. If a picture saves a thousand words, data could save millions of words. Within the art of data there may be many different disciplines: data sciences, data management and data visualisation, to name a few. The reason we call it the art of data is because of the overall level of creativity needed to solve the problem.

How does an organisation establish and sustain a data culture? In our last book we emphasised the importance of establishing the role of the CDO to deliver this outcome. This is less about putting a person in place but more about having a central focus to make sure that the data is treated like the asset it is at board level. An organisation that realises that it requires a CDO is an organisation that is stepping towards a data culture. Appointing a senior individual to be responsible and accountable for data, an essential asset to the business, is a clear signal both internally and externally that the company values its data and is building a data culture. In that book we go on to suggest that one of the primary tasks and responsibilities facing a CDO is building the data culture.

The leader of data literacy is the CDO. The chief finance officer (CFO) looks after the asset that is money, creates a culture where the business values that money and looks after it, and is able to tell the story from the P&L or budget forecast: the CDO has a similar role with data.

Data triangles

For an organisation to undergo data-driven business transformation it must understand its own data ecosystem. In Chapter 2 we discuss the data maturity model, why it is important and how to conduct a maturity assessment. Much of that process will reveal the data ecosystem within in an organisation. However, before we get into that level of detail it is worthwhile spending a little time understanding what a data ecosystem looks like within an organisation.

All companies, whether they know it or not, and whether they accept it or not, have a data ecosystem. The triangle is made up of people, technology, processes and, not very surprisingly, data. Each of these can be looked at in turn to understand its place and role in the data ecosystem. Fully understanding the data triangle will enable an organisation to carry out a data maturity assessment and embrace the potential of data-driven business transformation. It really helps to know where you are so that you know where you are starting from.

The technology in the data triangle may be quite complex, but we don't need to concern ourselves at this point with OT beyond understanding that unless data is gathered in a managed and governed manner, with an eye to the quality and end use of the data for business insight and decision making, then it is of little use to the data triangle. In this respect data validation and verification at the point of entry, or gathering, is vital to the health of the data ecosystem. The next step is to understand the flow or the communication of this data into safe storage. If the flow is not secure then the data may lose its integrity or volume and again will adversely affect the health of the triangle. Understanding these two fundamental elements of the technology in the triangle is therefore vitally important. Insight and subsequent decision making cannot be based on poor quality data, or even worse on data where the quality of the data is not known or understood.

The next step in the technology part of the data triangle is data storage. Where and how is the data being stored? Is it safe and secure? Is it stored in a place and manner that makes it accessible to the business? Referring back to Bordonaro, data must be easily accessible to the business. Understanding how data is made accessible to the data triangle is essential. We have heard of situations where organisations are dependent upon outsourced third parties to provide data sets that imposes time and cost constraints upon easy access. This situation makes data-driven business transformation very difficult.

The next component in the technology element of the triangle is the tools for extracting data for use within the business – the E in ETL (extraction, transformation and load). Are the extraction tools fit for purpose? Is it possible to extract the right data at the right time? If not then the power for data-driven transformation is diminished. At this point it may be worth mentioning that the data ecosystem may have a separation between the OT and the data and analytics technology. The extraction tools might be pulling data out of the OT to load into a data technology layer. Anyway, once the data is extracted what tools are available for data transformation and are these fit for the purpose of driving transformation of the business? We have all come across instances and businesses where the predominant, if not favoured, tools for data transformation are XLS and Microsoft Access. Often these tools are used by the business in the absence of any alternative or proper tooling. The use of these tools tends to lead to ungoverned data sets being use for analysis and business decision making, they may even be unsupported by the IT team. If data transformation in the current business state is undertaken in XLS or Access then business transformation driven by data may be very difficult.

It is important to understand the location of the technology element of the data triangle, is it on the premises or cloud based, or a hybrid of the two? Finally, what documentation exists around the technology element of the data triangle? Is there an architecture to the data in the OT and a mapped data lineage across the OT and IT applications and into reporting and analytics?

Moving onto the purer data element of the ecosystem: is the data governed, are its quality and standards understood, is it owned and managed, and if it is a yes to all of these who in the organisation does these things? Is there an information asset register? Within the data element what is the matrix of tools, platforms and coding languages being used to develop and deliver data products into the business? Are third parties being used at this stage in the ecosystem for data analytics? Part of the data ecosystem is made up of the many processes that are used to develop and deliver the data outputs and analytics and the layers of assurance and governance around these processes.

The people are a vital part of the data triangle. Who are the players in the data ecosystem, what skills do they have, where do they sit within the business? Who do they report to by task, line management and profession? Are the players set up in such a way within the triangle that their activity and data become siloed and separate?

Finally, within the data ecosystem is the business: what data and insight do they need and what are they getting? Where in the business is the data and insights being used and by who, what level of skill and data literacy do they have? What is the level of the data culture within the business? See Figure 1.7.

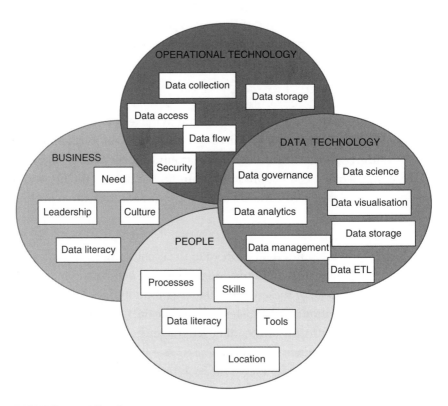

FIGURE 1.7 The data ecosystem

We call it the data ecosystem because each part interacts and the whole is diminished without any single piece. There exists a complex relationship within it, which when working together forms a whole where in some cases people don't even acknowledge the parts anymore and just see the whole enterprise. Not being able to see the wood for the trees is a great analogy here. What we are going to help you with is seeing the trees for a better wood.

Change adoption

Any business transformation, to be effective and deliver the end state, will require the transformation to be adopted. In practice transformation may be a series of smaller changes that collectively deliver the transformation. What's important is that they are all heading in the same direction, each of these individual changes need to be adopted.

The initial step to ensure transformation is to get 'your head right' at the start of the whole transformation process. This requires complete business buy-in for the transformation, the most senior sponsorship is imperative. Engage the organisation and prepare them for change. There are many books written on delivering 'transformation' and we don't wish to repeat those here. Suffice to say that to ensure a data-driven business transformation data should be placed close to the heart of this planning and strategy. The business must be involved in the transformation and lead it, rather than have it done to them.

The second step for successful adoption is to truly understand the differences between digital, technology and data transformation. Unless this is clearly understood at the beginning, and the inter-dependencies between these mapped and appreciated, adoption will fail. In fact, the transformation process will probably have failed even before it reaches an adoption phase.

The third stage is understanding the data ecosystem. How does it all fit together? This leads onto a data maturity assessment, which is discussed in Chapter 2. What these first three steps tell us is that proper preparation and engagement will increase the chances of adoption of the end state.

Part of the preparation leading into action is creating the right team to design, drive and deliver the transformation. This has to be a true blend of business, domain experts, technology and data. We would also include 'visionaries' in this broad blend of skills. Visionaries are people who can imagine the end state and can articulate and communicate this vision. They are as important as the technical unicorns. Adoption of the transformation will require changes to culture, changes to corporate vision, changing and persuading self-interest groups. It is the visionaries who can lead this. The visionaries may be the founders of a start-up, people who can imagine innovation and disruption, who can describe the future vision to investors and get them to invest. The visionaries may treat the stakeholder groups like investors or early adopters and sell them the vision.

Within each of these broad groups required in the integrated team there will be a need for a range of specialist skills. In the data group that may include data engineers, architects, analysts and data scientists.

To achieve adoption there has to be a proper approach to change. In simple terms, it has to be accepted that transformation will bring change. We must expect it to come up against protests of 'but we have always done it like this' – a phrase which is guaranteed to drive any rational change agent crazy.

Some thought should be given to the nature of the transformation to be delivered to improve the success of the adoption. In reality transformation comes in different flavours. There is innovation transformation, which will deliver an end state that is the same business model as the start state but provides new ways of doing the same things. This describes the vast majority of change transformations. This flavour of transformation is less difficult for an organisation to adopt. Disruptive transformation will change the business model, possibly even change the revenue streams and is far harder to gain adoption of. Nokia, for example, started off selling rubber boots and Shell originally used to import and sell shells (yes really). This type of transformation may be delivered as a spin off from the main part of the business. A good example is Vodafone (the company that won the licence to build Britain's first cellular telephone network), which was a spin off from Racal, a radar and electronics company founded in the 1950s.

This raises the question of whether you build the new alongside the old. Or do you consider an organisational wide roll out overnight on D-day. Perhaps the transformation is delivered as a patchwork across the organisation, going at varying paces in different parts of the business but eventually ending up with the whole quilt once the whole patchwork has been created. This is a question that needs to be bottomed out as part of the preparation.

The organisations that best adopt transformation are those that start the process expecting things to be different at the end of the process. They may not fully understand the exact shape of things to come but they are prepared to trust the process.

Transformation blockers

One of the biggest blockers to transformation is the failure of adoption, in some cases the fear of the failure of adoption is the drawback, and that is why we have dedicated the previous section to this. Apart from the failure of adoption we identify seven other blockers to transformation:

1. Not having a clear vision of end state.

2. Lack of adequate funding.

3. The wrong people.

4. The wrong skills.

5. No culture of transformation or no data culture.

6. Transformation is too hard for the organisation.

7. Disruption elsewhere: internal or external.

Not having a clear vision of the end state

If the end state vision is ambiguous or unclearly defined or understood, there will be faults and cracks in the preparatory stages and it will be all too easy for naysayers to block the transformation. A lack of clarity, or a failure to communicate the visions clearly, will be a constant blocker throughout the transformation process: every snag will turn from being a mole hill into a full-blown mountain. The vision is the guiding star: if the star is not clear and bright then blockers will appear from the shadows. Don't forget that narrating the data-driven transformation may be difficult, and requires specialist skills and experience.

Lack of adequate funding

A lack of proper funding will blow a hole in any change or transformation. It isn't necessarily a lack of funding in the case of a data-driven transformation that is important, it is the lack of appropriate funding. Data-driven transformation will take the business, and the procurement team, into new territory around licencing and procuring software and platforms. Also, if the data-driven transformation is being delivered as a series of agile iterative processes, perhaps part of a larger patchwork of adoption, there will be moments of discovery and pivots that may not fit rigid funding cycles and models.

The wrong people

All transformation is based on people, technology and processes (and data) as discussed earlier. It should therefore not be surprising to include people in the section on blockers. We deliberately haven't covered technology in the blockers, we will cover that in Chapter 8. Calling out people as a blocker – as opposed to skills, which comes next – is having people who are engaged, leaning into the transformation, 'change villains' or people who engage in 'change containment' are the wrong people. If they aren't with you then they are against you. The integrated transformation team must have the right people, people who can embrace change, understand change, understand data and communicate the end state with passion and vision. In a wider context the organisation that is transforming needs the right people who will adopt the transformation and engage with it.

The wrong skills

This again has two layers. The integrated transformation team needs the right mix of skills and experience. A data-driven business transformation will need people with data skills that may well be new to the organisation. So, the business will need to recruit the right skills, that is, the right people with the right skills. Knowing which skills to obtain will require specialist leadership.

The second layer is in the wider organisation. If a data-driven business transformation is to be delivered then it is very likely that the business will need upskilling in terms of data skills. This brings us back to the M&S example and data literacy.

There may also be a lack of IT skills: if the data-driven transformation is bringing in new thinking, new approaches and new technologies then other parts of the integrated team need to step up with the right skills. A simple example is the organisation that has run using on-site data centres and has an internal IT team or third party supplier team that is built and equipped to support the one location. A problem may arise with a lack of skill, experience and capability if the data-driven transformation is building an end state based on cloud platforms and software as a service.

No culture of transformation or no data culture

Lack of a culture for transformation and lack of a data culture will rapidly become blockers to successful data-driven transformation. Both of these may seem very obvious, but they are quite subtle. We have been involved with organisations that have 'talked' a great 'transformation plan' and spent time and resources on developing the plan, and even started to deliver it, but because the organisation is not set up, either through lack of leadership or middle tier resistance, the transformation plan has faded away and been quietly forgotten. How many times have we seen branded mouse mats, mugs or other items knocking about an organisation as an uneasy reminder of a long-forgotten transformation plan that was going to change the world? Some organisations become serial offenders in this respect, and often have a complete lack of corporate self-awareness or recollection at senior levels. The memory persists at lower levels in the organisation, embodied in the phrase 'oh we've tried this before and it didn't work'.

Transformation is too hard for the organisation

This is a difficult one to overcome and is driven by a couple of potential forces. First, the organisation is so busy surviving that there isn't any time or resources for transformation. The second is that they don't think that 'data' applies to them. This means getting their heads around, or even seeing the opportunities in, the data for them is too hard, a jump too far.

Disruption elsewhere: internal or external

The final blocker is disruption, which can be either external or internal. These can both be addressed in the same way. A transformation pro-gramme may be threatened by both external and internal disrupting forces that may make the transformation vision undesirable, outdated or not appropriate. There are literally some things that you just can't prepare for (usually they are a set of events that result in a swiss cheese type example of the perfect storm rather than one individual event: you tend to be able to prepare for one event, it's the cumulation that can surprise you). These are the hardest blockers to prepare for, by their very nature they are disruptors.

We also need to take a moment to talk about two very different types of characters that you will encounter along your data-driven transformation journey: the data villains and the data heroes.

We always like to start with the negative in order to finish on a high. So, let's cover the data villain first. They are those happy souls who do not want this transformation to happen, for whatever reason that is, and we will look at that in more depth in Chapter 6. Nevertheless, they are not on board with the programme. Even worse than that they may appear to be supportive while carrying out guerrilla activities in the background. This happened to us at one company in particular. A member of the executive board who should have been a supporter (in fact they had part responsibility for the data, which made their attitude all the more surprising) was openly supportive in the executive committee but spent a great deal of energy undermining any change activities. The executive in question held lots of one-on-one meetings, asking leading questions and generally setting out to negatively disrupt the data journey. This only came to light when one of the other board members decided that this had gone too far and raised it with us as a concern. What resulted from that was a great deal of completely unnecessary catch-up work to get the programme on track. When it was finally resolved, we determined that it had all been about personal politics, which was the most disappointing aspect of it. Finding out about such people is crucial so you can either change them into a more positive mindset or at least into a neutral one.

That said, the data heroes are your superstars, they are the people in the organisation who 'get' data. They probably have been banging their head against the company walls fighting the data battles in isolation, creating policies for their own areas, but may not have joined up because they didn't have the overarching view across the company to allow them to do so. These people give you your immediate business value, they are committed, passionate and a very useful resource. They will, however, likely need a little tlc at first, due to having been in their own isolated data wildernesses, but they can be the most powerful resource going forward.

Chapter 2

Understand Your Starting Point

W e have access to more data than ever before but it has become confusing and contradictory to the point that the quality of data is considered a discipline in its own right, as if it isn't a given you can trust it. At what point did we stop trusting data? It is not only the volume of data that makes it confusing it is how we use it.

To fall back on a very old quote from Mark Twain, 'There is something fascinating about science. One gets such wholesale returns of conjecture out of a trifling investment of fact'. Or to put it another way, we can use data to give us many different outcomes depending on what we do with it. If we don't stop to check that we are agreed as we manipulate or extrapolate from it then we are heading in very different and competing directions.

(Of course, Mark Twain also said 'Data is like garbage. You'd better know what you are going to do with it before you collect it'—and we are sure that there are lots of data scientists who would argue with that point, but maybe that is an argument for another day.)

This chapter is about knowing where you are starting from. The whole point of this is to make sure you understand where you are so that everything you are planning can be built on solid foundations and that you clearly see the picture you are dealing with. It's also about understanding where you are trying to get to. The start and end of the treasure map.

You are a data adventurer and you know there is a really valuable shipwreck somewhere in the ocean. This is you finding out where the ship is and understanding the terrain to map out how you get to your treasure. Only, in this case, you get to decide what you are going to try and acquire from your effort.

Remember this: the only stupid questions are the ones you don't ask! Don't be afraid to check understanding as well. Different people use the same terms to mean different things. We have worked on projects where major failures occurred because of simple misunderstandings between two people who believed they were talking about the same thing but in fact were miles apart. They hadn't taken the time to check that their basic understanding was the same. It wasn't a case of comparing apples and oranges but rather apples and orangutans.

We want to make people curious about the data they have, like a child's curiosity you never know what question it will lead to but it might turn up some really interesting answers to things you never knew you needed. Questions can drive the data journey forward. How can we tie it back into measurable outcomes (both good and bad) that enable us to stimulate the debate and take action in an agile flexible way?

You have a nice blank piece of paper, so where do you start? Well you don't actually have a blank piece of paper—that would be a luxury. What you do have are lots of things all competing for your attention, ongoing projects, data heroes and villains, politics, risks, issues, opportunities and a mountain of expectations to deal with. What you have is your legacy, be that infrastructure, attitude or landscape.

The first step should be doing a data maturity assessment to help you understand what the landscape is you are dealing with – it gives you a picture of where you are as an organisation and finds your starting point on the map. The data maturity assessment is an incredibly useful tool in your tool bag. Not only is it critical to the understanding of the overall state of the organisation before you start (or from when you get a chance to start the process) but it will also be useful when you are demonstrating progress as you will use this as a baseline in your ongoing assessments. Lots of changes fail because time wasn't taken to understand the true position of the company and moving your company to be more data enabled will mean you are making changes to it. A caterpillar can be a perfectly good caterpillar but to reach its true potential then it needs to turn into a butterfly. While we are not suggesting that it is possible to know everything in advance (its blatantly obvious that it isn't) we *are* suggesting that if you take a little time to understand where your company stands on the important factors in relation to how it treats data then it vastly increases the chances of you being successful.

It can also give you a baseline to demonstrate what you have been doing and where you have made a difference (and it helps you understand if it is going to be worth it!). One of the things that is hard when you do any kind of change is demonstrating the benefits – without that it is really easy for any company to worry that it is not on the right course. Nervous stakeholders and customers can take a great deal of energy to manage and may also derail programmes altogether if you aren't careful. A simple way of helping them understand that you are steering the ship in the right direction is to demonstrate progress against your baseline – which is what the first maturity assessment gives you. We will discuss how to take this forward later.

Some points to remember as you conduct your maturity model:

- Every interaction is a chance to engage. Thus, make sure you use them as such. Have your preparation work done so you understand their motivations and areas of concerns and how the data journey you are taking the company on will help them – no one in the organisation is immune to the 'what's in it for me?' conversation. This is a theme we will expand more as we progress through the book.

- Not everyone you meet will be a supporter or even mildly positive about what you are trying to do. Use this time to better understand where they are coming from so you can decide if you want to put the effort into changing their position.

- If they are positive don't assume they will stay that way, continue to engage making sure you maintain a good level of buy-in.

- Take every opportunity to change people's attitude to data!

The main elements of the maturity model are:

- Strategy – mature organisations communicate the vision. That is, leadership's way forward with a business strategy to inform and provide the principles for detailed strategies relating to key areas of the business, of which your data strategy should be one.

- Corporate governance – are the key elements of good corporate governance in place and are they well deployed? Do they operate in isolation to each other or ensure a well-rounded approach is taken? Are relevant and tailorable assurance activities regularly undertaken and well used? Governance should cover assurance and compliance as well.

- Leadership and sponsorship – are there key people at senior levels within the organisation who understand the importance and value of what you do? Will they back you to make sure you get the time you need in order to demonstrate the true value you are bringing.

- Framework, process and tools – does your organisation have the right framework in place to make the rest of your pieces hang together? How do you bring the policies up to date and make sure you keep them that way? Do the tools you have access to help or hinder you? How many data-related systems are you currently using and are people using them for the right things? Do you understand the information lifecycle within your organisation?

- Policies – do you have the relevant policies, standards, procedures and so on to make sure you are setting up the people within your organisation to succeed? Are your instructions clear, consistent and easy to use? Do you have a framework in place to demonstrate the interrelation between your policies?

- Information risk – is your information risk well-defined and at what level? Do you understand your business criticality? Are the tools in place to help you manage and mitigate this appropriately?

- Architecture – do you understand how you use information across your organisation, how it cuts across your silos? Have you mapped your architectures out? Who is accountable for which bits?

- Organisation, roles and responsibility – are roles clear and agreed across the organisation? Do you have a team dedicated to being the data cheerleader for the organisation? Have roles been defined to address elements of information management and assurance within your different domains. Do you have a steering board in place that is empowered to make your data and information-related decisions?

- Skills – why type and level of skills already exist for you to work with in relation to where the organisation needs to get to? Are training programmes in place for both your data and information profession- als and for the wider organisation?

- Metrics – firstly, are you measuring progress and performance mea- surements, reporting or benefits capture at a corporate level. Then are you measuring the right things to drive the kinds of behaviour that you want in your data-valuing culture?

- Information guardianship behaviour – do you value your data and understand what it can do for you? One indication of this is how much money you have spent on it and over what time period. Has it been underinvested or has it been well invested but not succeeded, and why not?

- Technology – technology has massive legacy costs associated with it, some of which may not be easily adjusted, or, when you start dig- ging, multiple versions and types of software may be confusing your current workforce.

The other important aspect of your maturity model is the levels that you would 'mark' yourself against.

   0 – Unaware, we know that normally you don't have a 0 but you would be naïve to not understand that in some cases organisations aren't even aware they have to think about something – yep we have all

been there, the old unknown unknowns! Spreadsheets everywhere, it's where Excel rules! Everyone is taking data from all over the place and modifying it without any thought to what other people need it for – basically the data equivalent of the Wild West.

1 – Aware, few processes other than those required by legal demands and any industry regulations, ad hoc efforts with very low levels of trust.

2 – Reactive, some processes are emerging, not monitored and regularly worked around, data quality checks are done ad hoc, the idea of standards are just starting to emerge but normally in response to an issue faced or direct data breach.

3 – Proactive, processes developed by information stewards but with limited authority: things like master data management programmes are started and governance of unstructured data begins to emerge. Starting to get ahead of the curve but there still seems like a long way to go.

4 – Managed, information governance organisation is in place, functions at enterprise level, stewards and champions are working together and things are evolving into a proper framework, which is communicated and assured.

5 – Optimised, nearly all information assets are inventoried, including knowing external sources, information advocated across the business, enablement and value generation. Transformative data is used internally and externally to make sure the organisation is cooking with gas. True partnerships with customers and supplier are embedded and the data landscape is trusted and agile.

What you should be looking to end up with is a table that looks like Table 2.1.

Well maybe not this one exactly – it's not a great score but if this is your starting point be realistic about it.

TABLE 2.1   Example of a data maturity model

Element	Score	Element	Score
Strategy	2	Architecture	1
Corporate governance	1	Organisation	1
Leadership	2	Skills	0
Framework	0	Metric	1
Policies	1	Behaviour	0
Risk	2	Technology	1

TABLE 2.2   Amalgamated data maturity model

Element	Score	Element	Score
Strategy	2.06	Architecture	1.93
Corporate governance	1.9	Organisation	1.46
Leadership	2.2	Skills	1.6
Framework	1.6	Metric	1.53
Policies	2.46	Behaviour	1.8
Risk	2.4	Technology	1.86

We have conducted research with companies across the world and the average company maturity model looked like Table 2.2.

As you can see from this picture, we have a way to go to say we are getting the most from our data. It doesn't matter where you start from, what counts is that you take the first step, then the next and the next one after that. A tortoise who keeps going will always pass a sleeping hare.

We know that it would be great if we handed you a list of questions to use, so we have included example questions in each section. If you chose to use the example questions in that way then you are welcome to. However, we are big fans of one size does **not** fit all. The main point of the maturity assessment isn't so that you can compare yourselves with every other company, it is so that you gain a true understanding of where you are and the progress you are making. Therefore, it could well be that some questions are more important to you while others have no relevance at all. When models are used it is too easy to fall back on squashing your organisation into the model rather than understanding the point of it and adapting the model to really help you.

The other thing to look at is whether any of the areas can be merged, we have made sure that every important element is included and they should all be covered; however, if you have a mature organisation who really understands its overarching governance we might be tempted to merge the policies with the framework section. Alternatively, if you instinctively know you are in a really bad place, step through each element really carefully. We aren't big fans of being prescriptive and don't think you should be either as every company is subtly different.

Let's take some time to go through each element of the maturity model in turn to understand what you are looking at and the type of questions you would ask within each element.

Strategy

Mature organisations communicate the leadership's way forward with a business strategy to inform and provide the principles for detailed strategies relating to key areas of the business, of which your data strategy should be one.

First things first, before you even think about putting a data strategy in place (and make no mistake you need a data strategy) make sure you know what the business strategy is. What direction is the organisation heading in, who are your main competitors, what are your immediate business threats, what are you currently doing and what do you want to change? This is where the key to data enabled rather than data driven comes into play. If the organisation is a start-up then data driven might be the right term to use. However, in the majority of cases you are trying to change an organisation that is already in a state of 'business as usual' so enabled is the better word to use. Your data underpins the company, and the company already exists, so what do you want to do differently, how is your data going to ENABLE this change?

What you are trying to understand with this element is how mature your data strategy is and whether it contains all the parts that make sense for your organisation. If you are a traditional company that has many complex parts, does it take that into account? Or, if you are more in start-up mode, does it talk about slowly introducing the governance to take you to a more mature organisation?

If you are really early in your journey, is your strategy part of a programme mandate rather than standing alone at the moment? Don't leave any stone unturned and for goodness sakes don't assume. We have been presented with a resolute yes to there being a business strategy only to find out when we asked to see it that it didn't exist outside the head of the person we were talking to.

Tie in with your business strategy, and as a senior leader in your business, if your business doesn't have one, then do the grown-up thing and promote the idea of having one, help to write it, do what you need to do to get people on the same page as to what your company wants to be. You can have the best data strategy in the world, but if you don't know what it is that your company is trying to achieve you will still end up going around in circles. They might be fast and beautiful circles but you will nevertheless still be spinning on your tail. Sitting back and moaning won't solve the problem.

One of the very next things to understand when talking about strategy is that you don't have to fix it in stone right at the very beginning. In fact, we recommend that you don't. You should not assume that you know it all at the beginning just because you are pointing the organisation in the right direction. You might need to refine the direction everyone is heading in. Do you have an immediate data strategy that will move to a longer-term one or a way of keeping your data strategy updated? Have you built in the flexibility needed to allow and even encourage your business to evolve?

Understanding where on the risk adverse verses value add scale the organisation is and what it needs is essential – they aren't always the same thing, in fact, they are very often not. Companies are particularly forward facing and are looking to maximise profit. This is not a bad thing, it keeps people in their jobs. However, it does mean that, when asked what they want, staff will tend to focus on things that will bring more money in or make them look better. It's not as fashionable but remembering that if you save money you also 'make money' can be useful too. Henry Ford's statement, 'if I had asked people what they wanted they would have said faster horses', is often used when we want to really spur on innovation, but we believe that, if you look at it as simply as possible, the problem he was trying to solve was how to get from a to b. That's a different question and the question you ask may well elicit very different answers – so boil your strategy down to asking the right questions. Ford's phrase is one we use a lot but one that will genuinely help you to get your basics right. Your data house won't last long if it doesn't stand on a firm foundation – so does your data strategy take account of the risk adverse stuff as well as the value add? Have you an idea of the quick wins to focus on that help demonstrate the value you will be delivering in the longer term? At this point it really is about balance. Don't let your company be pulled too far in any one of these directions at the beginning. But when do you think you are ready to really drive the value, this should also be in the strategy.

If it is important to you, have you thought about how you will work with suppliers and partners? How does your company feel about insourcing or outsourcing, what is the current landscape with this regard and what does the future hold? There are lots of great initiatives and partnerships happening between businesses, academia and governments. Is this something that your strategy addresses?

There is no point in having a strategy that is at complete odds with what a company believes to be true. If they have had a terrible experience with outsourcing in the past, for example, then there is no point having a strategy that has this at its heart. The resistance that will happen as a result will be disproportionate to the effort needed to resolve it. Even if that is genuinely the right thing to do, putting it as a lynchpin doesn't help the overall cause. Timing is everything here.

Does your strategy look at what is happening externally? There is always some regulation or external governance that needs to be thought about, but what about wider disruptions in the market, have any of those been addressed? Are there threats or opportunities over the horizon that you need to keep an eye on? Think about areas of complexity, distribution, diversity and scale – what are you declaring you are going to integrate and maximise the value from?

Now that is an awful lot to think about in a strategy and if you try and cover it all in one document you will have probably rewritten the works of Shakespeare – a bit weighty to be inspirational for your company. Hence why you need to look at what the important parts are to your company. Prioritise what is more important for you and what helps you form the guiding light for the organisation.

It's all about creating the story and linking it to the golden thread – what role will each person play to really drive your company forward. Every part of a car is important but it all has to work together in order for the car to move. Does every part of your car understand their function and how it contributes to the overall machine?

Which brings us to another key understanding point – there is no point having the best strategy in the world if it sits in someone head or at the back of an electronic book shelf. Has it been communicated? Even better, has everyone engaged with it and what demonstration of this do you have?

Strategy

Is your organisation clear on how it treats data?	Does a data strategy exist?
Is this clearly communicated?	Does it clearly point out how the company will treat its data?
Are you enabling the rest of the company to be better?	Is it time bound?
	Does the data strategy underpin the business strategy?
	Is the data strategy clearly communicated to the organisation?
	Is it regularly reviewed?

Corporate governance

Are the key elements of good corporate governance in place and are they well deployed? Do they operate in isolation from each other or ensure a well-rounded approach is taken? Are relevant and tailorable assurance activities regularly undertaken and well used?

The reason why you need to get this right at the very beginning is simple, it's how you are going to make your key decision on the rest of the stuff you do. People hear the words governance and tend to focus on the negative connotations of it. They only see how it slows them down or stops them doing something. Now let's be clear, usually there is a really good reason why you shouldn't do something – big companies have been brought down because some staff believed that the rules didn't apply to them. Enron has to be one of the best examples of this. Enron failed because their reported financial condition was exposed to be as a result of institutional and systematic accounting fraud. This fraud also had a massive impact on their auditors Arthur Andersen (who split off their consulting wing, as it wasn't part of the scandal, to become Accenture).

As for slowing things down, in some cases it is true that going through a governance process can add an overhead on the time and effort you need to do something. However, let's looks at the positive side of things. By going through a governance process you have to understand what you are doing to a high degree. It makes you cover your bases, dot your 'i's and cross your 't's. We know that there have been cases where going through that process has made a project change tack because certain aspects hadn't been taken into consideration, such as information security considerations, for instance. It's good to have a check point. Anyone else chant 'tickets, money and passports' before they leave the house to go on holiday? We use checks all the time in our everyday lives to make absolutely sure we aren't missing something. Think about checking that the cooker is off or the door is locked – just in case. Well, corporate governance is a company's way of checking – just in case.

The other side of governance is that it puts everyone on a level playing field. We have all been in a situation where one person gets their way through sheer force of their personality or because they are quite simply a bully and have trained people to live in fear of them. A proper governance system minimises the possibility for this to happen. Nothing totally eradicates it but it limits options for someone who is just trying to get their own way. (They may genuinely believe that what they want is the right thing for the organisation as well, fighting for what they believe in doesn't make them bad people, but it can make people blinkered – so focused on what they can see that they don't see the tank coming from left field.)

This is another non-negotiable when it comes to your maturity model: if there is absolutely no corporate governance then you have a problem. (We would surmise that every company has a degree of governance even if it is just you running your ideas past the boss.) If you are working in an organisation of more than five people there will be governance of sorts. It is much easier to tap into the governance that already exists than trying to impose a totally new model that will appear quite alien.

How does this tie into the rest of the corporate governance? If you are planting a garden with some well-established shrubs you minimise disturbance to good plants and clear away the dross, putting the new in the right place in relation to the established. This is what you must do here. You shouldn't be setting up a whole new set of corporate governance or running off declaring independence but rather tapping into what already exists and evolving it to uncover data governance as well. This will involve different streams but should feel quite natural when you are finished.

Defining what already exists is really helpful, but you don't need to just focus on what data governance exists. Really you need to understand how mature your organisation's corporate governance is as your primary action, because if this is something that is taken really seriously and actions happen as a result of the decisions taken then you have something you can work with. Having any kind of governance around the data is just a major added bonus at this point.

It is also a relatively easy area in which to move from level 0 on the model to a level 2 by putting basic governance in place. There are very tangible outputs to demonstrate what you have done, which help show that all important early progress when you are still in the building phase.

One of the key areas to focus on in this area is what governance already exists. Think risk framework and decision-making meetings (in fact any meeting where you can access money tends to be part of corporate governance) and then how embedded and joined up all these things are. We have seen great risk frameworks, award winning in fact, but they weren't fully joined into the heartbeat of the company so didn't make the impact they were capable of. If you have an overarching corporate governance structure, which is clearly defined and the organisation buys into, firstly rub your hands with glee at your good luck and then go find the person who oversees this as you are going to want to be best friends with them.

Another thing to look for is whether data governance is wrapped in (or assumed to be wrapped in) with the governance surrounding IT. Is this the only place it is mentioned? That is better than nothing but also a warning in its own right that you have to work on separating the idea of data from technology in your company's mind. It's like assuming that the driver and the mechanic have the same skills in car racing. Both are critical skills but very different.

What interplay is there with the audit department, are they also part of this process and are they open to working with you? Assurance is a large part of making governance work and is a more comfortable way to engage with the business (more on this in Chapter 3) make time to explore how the organisation feels about this fit. Is there already a contentious relationship that exists when you mention assurance or is there an open door?

Essentially you are looking in to the state of the operating model. Rather than being prescriptive and suggesting one model, there are many different models – and pros and cons of each one. The model you chose is dependent on which one fits your organisation and which one you are comfortable with. However, it should be something that everyone understands and buys into. An operating model that looks good on paper but never leaves the page isn't worth the paper it's printed on. If there is any ongoing work looking at a new target operating model then this is a great opportunity to embed your governance as part of that process. Since the target operating model provides the vision for organisation to change that is definitely something you would want to know about and capitalise on.

This is a little aside but is wrapped up with the operating model. Having a view of the company's value chain also helps you later when you are targeting where to focus your energy.

Finally, in any game (and yes business is a type of game, everyone is trying to win by constantly striving for an end goal, it's just on a different scale to those games of monopoly or trivial pursuit you play with the family on a raining Sunday afternoon) you establish the rules before you play. Otherwise you end up playing a giant game of Calvin Ball, which while it looks like fun at first always descends into a giant show of chaos wrapped up in pandemonium.

Understanding the rules – everyone playing the same game – allows teams to have strategies and work together; and it gives you a way of keeping people on track. Well your governance is the rules and referee of the game. Someone will always try something radical but that just spoils the game for everyone else. Once you get really good governance embedded, it is self-tracking so that the whole team understands the need to for it and buys into it so that the system becomes self-correcting.

Corporate governance	
How mature is good governance within your organisation?	Does a governance framework exist?
Do the different elements of governance work well together?	Is it effective?
How can you demonstrate it works?	Do any governance elements sit outside this framework?
	Do assurance activities take place?
	Are actions taken as a result of the assurance results?

Leadership and sponsorship

Are there key people at senior levels within the organisation who understand the importance and value of what you do?

Will they back you to make sure you get the time you need in order to demonstrate the true value you are bringing.

We can honestly say from personal experience that if you don't have the right leadership you will have a very difficult time indeed. It's not just about having one person at the top of the tree who everyone will defer to and comply with (although if you do have this – don't knock it!). It's about the leadership of the company believing in the change you are driving, not necessarily everyone (you can't please everyone all of the time). However, you need the majority on your side, or at least ambivalent. Part of your role in the getting ready phase is getting into that position. Remember all that time you invest in people will come back three-fold – the senior leadership of your organisation are no exception.

The counter to that is also true: with the right leadership in place who 'get' what you are doing then you've been given a major step forward. One example we have is of working with a chief executive who bought into a vision so thoroughly that they refused to deviate from the policy themselves even though (for a short time before it bedded in) it meant extra work for them. It also meant that those around them had no choice but to follow suit. The ripple effect right through the organisation meant the policy was implemented and then successfully embedded in record time. Now no one questions why they ever did it any other way.

You will be utilising your stakeholder management skills here like no other time. They all have to believe in the direction you are taking the organisation and want to invest effort into helping it happen. This means that you have to make it clear what's in it for them. The greater good is a very noble idea but it won't excite a finance director as much as how much money you are going to save on an annual basis (just demonstrating that you will save more than you cost in the first year is a good start while you work on your quick wins and longer-term strategy). Getting information to the front line that frees them from waiting on decisions from on high and enables the rest of the department to focus on more than the minutia will get you buy-in from an operations department, for example.

Don't settle for just having one major stakeholder either. They can be overwhelmed or out voted if the rest of your board are detractors. This should be such a positive message that you have the majority of the board on your side, or at the very least positively neutral to give you an open platform to have the right conversations without prejudice. It makes their lives easier as well if they aren't constantly having to fight battles on your (the data's) behalf.

Be honest when you are answering questions about the leaders in your organisation. It can be awkward, especially if they are going to see this but it does come down to priorities and at the very least give you a starting point for the conversations you need to have. Do their actions match their words? If push came to shove would they back another programme over what you are doing? There is nothing wrong with that, business have lots of different priorities constantly competing for attention. You just have to make sure you are aware of where the data sits in this sea of noise.

Quite simply the leadership element can't be overlooked at any stage – it gives a solid foundation for when things get difficult, before you really start on the doing phase. Get your ducks lined up. Know who your supporters are, who your detractors are and who is open to conversion.

Since what we do is driven by business outcomes it should be the business that gives you the mandate and supports you with it. You might just need to educate them on how fundamental this is to everything else they are doing.

Leadership	
Are there key people within your organisation at senior levels who understand the importance of managing data appropriately? Will they back you? Will they work together with data as the priority?	Is there a senior sponsor? Do they sit on the executive board? Does the executive committee as a whole buy into the data strategy? What relative priority does the data journey have against other strategic initiatives? Has clear funding been agreed to over an appropriate timeline? How stable is the political landscape?

Framework, process and tools

Does your organisation have the right framework in place to make the rest of your pieces hang together?
How do you bring the policies up to date and make sure you keep them that way?
Do the tools you have access to help or hinder you?
How many data-related systems are you currently using and are people using them for the right thing?
Do you understand the information lifecycle within your organisation?

The framework is how all your data space hangs together. Essentially, what decisions do you need made on a regular basis, what spans of decision making do you need in place and how does it all work with other decision-making bodies? Don't overload the organisation with unnecessary meetings – no one will thank you for that. Despite the running joke that meetings are the practical alternative to work, they can actually be a brilliant hiding place for anyone who wants a little break from doing something useful. That said, it is vital to make sure you have the ability to discuss important issues at the right level and most importantly actually make decisions – no company needs more meetings to just talk about things. If there isn't a specific reason for a meeting – don't have it! Every meeting should have a point to it.

So, effectively, you are looking to check if you have the right people in the right place to make the right decisions in a timely fashion.

Are there different levels of meetings to make sure the right thing is discussed at the right level? If you have really strategic people discussing an issue in detail, then they will get bored really fast and not turn up. If you have very detailed people discussing strategic decisions for which they can't influence the budget, then they will lose interest quickly and fail to show up. It's important to pitch at the right level. Make sure you know who is capable of making which decisions and who is empowered to make them – they may be different people and you need them to be the same as much as possible. You also want each of them to feel useful and that they have a part to play.

At the very least is there has to be an overarching meeting to make sure the whole business is joined up. There has to be something that means that each part of the organisation has a chance to discuss and agree on the really big stuff that has an impact right across the company.

Another area that it is imperative to cover is how the information architecture impacts the enterprise architecture and vice versa – a link between what is happening with data and what is happening with it to make sure the two interests start to converge is helpful. The other area to check is that the governance documents are covered as well. How are your policies kept up to date and are there regular checks to make sure they stay up to date? As well as the individual meetings being in place, is there a link between them, do they all work together in some kind of coordination to make sure that everything is working as a well-oiled machine rather than individual squeaky toys?

This is all about understanding the heartbeat of your data-related organisation, seeing the ebb and flow of information, where it's working and where it is being blocked.

It's worth knowing that not only are the meetings in place with a clear purpose but that people who are supposed to attend do attend rather than send delegate after delegate. If they are not present, that's a really good indication that they aren't getting what they need and you're in danger of losing an ally. People make time for the things that are important to them.

Framework

Do all the different places of your data organisation fit together and work together well?	Do you have a data target operating model?
Do you keep everything up to date smoothly or is it more ad hoc?	Is it linked to the company operating model?
Do you understand the information lifecycle within your organisation?	Are there clear, documented ways of making decisions about your data?
	Are they embedded and bought into?
	Does each meeting have a clear term of reference?
	Do the different meetings work together well?

Policies

Do you have the relevant policies, standards, procedures and so on to make sure you are setting up the people within your organisation to succeed?

Are your instructions clear, consistent and easy to follow?
Do you have a framework in place to demonstrate the interrelation between your policies?

There is a clear link between the framework and the governance documents so often your questions in these two elements can feed each other. Don't be frightened of overlap but rather allow each section to work together and help with your probing questions.

We've used the term policies but really this should cover all control documents, such as all the standards, procedures, guidelines and whatever else your company uses to make sure it gives clear instructions to everyone who works with them and around them. The first thing to check is everyone's understanding about what these documents are for. We have worked in organisations where some parts used policies as the main document but in other areas anything that was called a policy was ignored and only standards were looked at. Even check if there is a definition of what they believe each of those documents to be. You can identify problems by asking some really simple questions across the company on their basic understanding about what a policy is.

What is the volume of these documents, have the company recreated *War and Peace*? Often it seems easy to make instructions clearer by just adding more and more detail but in fact this just causes more confusion. No document can account for every single situation and by trying to do so you take away from empowering people to use their own decision-making abilities. This causes problems when they hit something you haven't covered in the document. It also makes it much harder to keep the documents consistent with each other as all that detail makes it too easy to get lost in them. The other way long documents don't help is that no one actually reads them. Anything longer than five pages tends to turn people off and attention can easily wander. Our personal preference is for short tailored documents that you can drill down into to get the level of detail you need – however, this definitely needs to be looked at within the context of your organisation.

In the main these documents should be clear, consistent and easy to understand. Depending on the type of company you are looking at the governance documents will be structured in different fashions, so check they are the right ones for you. Any critical safety areas will by necessity be incredibly detailed but it can be tempting to treat everything in the company in this way and you don't necessarily need this level of detail if you are looking at who orders the milk for the kitchen, for example. So, checking the appropriateness of what you have is important.

Simple questions about the volume of documents you have can only tell you so much, it's also important which documents you have in place. There are some basic ones that you need to watch out for, such as your information security, retention and data classification policies. Look for any legislation that directly affects your organisation's data and make sure that there is clear guidance around those. This would definitely be an area of concern if you don't have these in place **and being used**.

You can have the greatest documents in the world but if no one can get to them then that doesn't really help you and you have just wasted a bunch of effort. Is there any kind of central repository or index of where all the control documents are? Are any areas repeated or overlapped? This tends to happen if the documents aren't stored together or an overarching understanding of where they are stored is not kept up to date. People are helpful, they try and do a good job. So, if they can't find something they think the company needs in lots of cases they recreate them. This is why documents are rewritten and policies are redone, usually in ways that don't match the original (if anyone can agree what the original was in the first place).

The other questions to ask pertain to whether or not anyone ever looks at them. Lots of knowledge within organisations is tacit knowledge that sits in people's heads. This provides a great excuse for answering 'that's the way we always do that' when questioned, which usually means that the employee's predecessor told them what to do and they just carried on doing that rather than question things or check the policies around it. Having the documents accessible is one part of the equation, the other is checking if anyone is looking at them.

The simple way of describing policies and their importance is as the 'rules of the game'. Hence, understanding them makes everyone's life easier. It aids with interaction between individuals and teams across the company.

Finally, check how long it has been since the policies were reviewed (not updated as they may not have needed to be) so at least you know they have been evaluated. Are they kept up to date? Gaining control of this area won't solve all your problems overnight but it will stop them from getting any worse while you work to fix the big stuff.

Policies	
This covers all data control documents such as policies, procedures, guidelines and standards.	How many control documents do you have in place?
Are you setting expectations clearly and setting everyone up to succeed?	Is this appropriate for your company?
Does the organisation buy into the policies?	Do they all fit together?
	Do you understand how they are all linked together?
	Are they easy to understand, simple and short?
	Are they embedded?

Information risk

**Is your information risk well-defined and
at what level?
Do you understand your business criticality?
Are the tools in place to help you manage and
mitigate this appropriately?**

Start by looking at what is currently there, normally pockets and overlapping risks. It's a bit like the policies, when what you need is an enterprise-level risk that deals with the overarching issue facing the organisation, whether it is risk adverse focused or doesn't have enough focus on value and then the other risks are linked to it and streamlined.

This is one of those areas that often either gets missed or has too little attention paid to it. In some cases everyone has had a go and you end up with a plethora of documents covering minute bits of the risks around data and information. Risk can feel like the boring end of the scale when all you want to do is deliver loads of value and concentrate on the mind-fizzling, heart-pumping stuff that excited you. Just think though, when you are doing the exciting boggling work do you really want to take the time then to cope with being blindsided by something you could have been looking out for and already stopped in its tracks? Never underestimate the power of a bit of preplanning.

Ideally, with risks you should have clear, well thought out risks that are structured in a similar way, consistent and match the hierarchy of attention you want paid to them. Having a tiered level of risks with one big hairy one that you talk to your executive board about is great but open it out as you work through the organisation and break it down into more relevant risks to the different areas they cover. This ensures that you have the right level of support for the right topics. Pretending to everyone who is senior in your organisation that everything in the garden is rosy and there are no weeds trying to break out will either make them suspicious or not equip them to deal with things in the appropriate way. Talking to people in a consistent, realistic and non-sensationalist way about risks shows that they aren't there just to cause wide-spread panic but as an important business tool and a fundamental part of the operating model.

A starting point for this should always be to ask whether you have any risks that mention data or information at all and, if you do, how many do you have? This first answer will give you a good indication of where you sit on the scale of risk importance. Rather obviously, finding out what level these are discussed at helps you also understand the importance that the organisation places on data and will help you look after what you value.

This is a great time to get under the hood and look at the evidence, get examples of the risks if they exist and use them to back up the discussions in the workshops. The evidence here is also valuable if it doesn't back up what is being said in the workshops. For instance, look out for a raft of people telling you that there are no policies covering data when in fact you can find proof that they exist. What is the underlying reason for this? Is it because the policies are too abundant so they overwhelm people? Can they not find them or do they just not know they exist in the first place? A little background work on this area before you walk into the workshops can ensure that you have a really intelligent debate.

Are the risks structured so that you understand potential causes and outcomes? What does the risk really mean to the organisation? Is the risk really bringing to life what could happen but focusing on the practical ways of mitigating that? Another area to check is the metrics around the risks: are there metrics and are there leading and lagging indicators? Is there anything that gives you a heads up that the risk may be moving into play so you need to up your game when it comes to dealing with it? One incredibly important aspect is to also look at the ownership of the risk.

As well as looking at the risks themselves, look at where and how they are stored – accessibility can be a key factor in where a company uses its risks or not. How do they work together? Just like policies, the risks need to be consistent with each other and while it is possible that different areas will have nuances within them, this doesn't detract from the high-level commonality in different tracks or levels for those specific areas. Acknowledge any areas of difference, don't assume you can pretend that they don't exist.

Risk	
Are your data and information risks well-defined and understood?	Do you have a set of data-related risks?
Does your business understand the criticality of these risks?	Is there a lead risk that centres the other risks?
Are you supported in dealing with these risks?	Are the interrelationships between the risks understood?
	Are the risks managed at the appropriate level?
	Are actions to manage the risks taking place?
	Are the risks regularly reviewed?

Architecture

Do you understand how you use information across your organisation, how it cuts across your silos?
Have you mapped your architectures out?
Who is accountable for which bits?

Some organisations believe that they already have this sorted but it is easy to confuse data architecture as defined by TOGAF®, which is part of the enterprise architecture framework for information architecture – that itself is concerned with the information map of your organisation. Originally information architecture came from website design, organising the structure of websites. It has moved on now to look at how information flows through an organisation rather than what data sits in what IT system.

When it comes to architecture you are looking for an understanding of where your data is coming from, what happens to it on its journey through your company and where it ends up. What data objects do you have and how do you define them as they become information, knowledge and hopefully wisdom? When do these stages change and what causes the metamorphosis? Basically, what path do they take, and can you spot dead ends (we would put money on you finding reports that have been created at great expense to the company for years and end up disappearing into the ether as the need for them has long since ceased to exist). Where do you have multiple data objects that at face value appear to be the same but in reality turn out to be totally different things? If you take the same data source and perform two different operations with it then you end up with two different objects. Has the company recognised this or is it still trying to claim it's the same thing?

What you are looking for is a balance between all the details and still being able to operate at a manageable level. Your information landscape can be complex and confusing and will always continually evolve so you don't want your organisation to have this locked down. You need it to work at a level where you can still make useful decisions but that allows the company the flexibility to be agile. Look at the consistency of the data not just the quality as, despite being two different areas, they can be confused as the same thing. Think granularity, distribution, latency and physicality of data in order to harness its power and really get it working for you.

It is all well and good having this in place (that would be an amazing start – if you already have that then give your company a pat on the back). However, even if you have it all in place, do you have a process for making sure that you look after it? Change is guaranteed so making sure you allow and even encourage that change is fundamental to this stage. Does it feed into the other related processes such as knowledge management or the quality management process? This should work nicely with your questions in your framework section.

It's worth making sure you know where you are buying data from. Companies now purchase data from different sources, and it's not unknown for different parts of an organisation to have agreements in place with the same companies to buy data (in some cases the exact same data a different department is already paying for). If this is the case you can save some pretty eyewatering amounts of money by consolidating these contracts.

A little look at the providers of data is worth considering but this can overlap with the organisation section depending upon how you ask the question. As long as you get the understanding, it's not terribly important which section you consider this in. What is important is that you do consider it. Who provides your data? Where is it coming from – both internal and external? What are their roles and responsibilities and are these well-defined all the way through to why and where they are in the overall lifecycle of data? Where are your main interfaces with the data and are these controlled so that we know what is really going on? What are the points where changes to your data ownership occur and does this relate to the changes in state (i.e., from data to information and beyond). Has your data lineage been mapped and understood?

Last but not least, what ongoing projects are likely to impact any change on data areas within the company? Usually these sit within the IT area but check for anything around policy changes or operations as well. A changing role can have a big impact on changes to data.

Architecture

Do you understand how you use the data throughout your organisation?	Does an information architecture exist?
Do you understand how the data and information flows through your company?	Are data domains clearly understood?

Are accountabilities clear and accepted?	Are all data sources understood?
	Is the data lineage tracked?
	Are data domain owners identified?
	Is it regularly reviewed?
	Is there a managed information asset register?
	How advanced are your data assurance activities?

Organisation, roles and responsibility

Are roles clear and agreed across the organisation?
Do you have a team dedicated to being the data cheerleader for the company?
Have roles been defined to address elements of information management and assurance within your different domains?
Do you have a steering board in place that is empowered to make your data-related and information-related decisions?

There are many different types of teams that can be used to look after data. If any type of data focus does exist then it doesn't matter what it looks like, whether it is centralised or federated, it could be classed as a centre of excellence or a matrix team. What is really important is if it exists in the first place and is right for the organisation. If you have a great model based on bringing all the data people together under one heading but your company operates well with small centres of excellence, then you will be pushing a boulder uphill, the same goes for the converse.

Another area to cover is whether or not there been any evolution of the organisation? Is the team new or has is continued to modify and shift to meet changing needs?

Are the names meaningful and do they actually do what their job description says they do? Are they clear for both the individuals who are performing those roles but also for the rest of the organisation? Sometimes it's clear that the people doing the job are absolutely clear on what they are doing but it's difficult for the organisation to see how those roles do or should interact with the rest of the company. It's about the whole picture and how it all fits together not just the small individual bits. Do the individual parts make sense as a whole? A RACI (responsible, accountable, consulting and informed) model can be used to describe the participation by various roles in completing tasks or deliverables. This is particularly useful for clarifying roles and responsibilities in cross-functional processes.

Bearing in mind what the direction of ambition for the company is, check that each 'tower' of what you are after is covered. Do you have tonnes of data scientists but no one covering governance? Is there any-one looking at policies or is there any representation from the business? Try and look at the totality of what exists. There is a danger that different areas can differ widely in this section as they are not focused on the same things. Some organisations can be seen as savants in that they have amaz-ing analytics functions but without spending time making sure there is a high degree of data quality going into it.

There's an opportunity in this area particularly to look out for. Usually there are a number of people who have been fighting the good fight for ages; however, because they haven't been joined up it has been very dif-ficult for them to make any head way (in some cases they turn into data villains and can end up doing more harm than good). This is a time when you need to look out for any of these data heroes lurking in the depths of the company as they are a really useful source of information and ideas. There could well be some of your early quick wins in the projects they have been working on in isolation. By helping them link up, or by giv-ing them the support they need, you can have wins for the organisation in unbelievably quick time scales because you aren't starting from scratch (just remember to give credit where it is due, they could be brilliant advocates or even become part of your team).

Is there any collaboration between the different data-related teams, does the information security team know what is happening with the data engineers, or are the decision scientists linked up with the governance team? You want to create communities across your business that can be connected and collaborate; however, they need to have enough information to be able to operate on their own when they need to. So, do they operate together out or good will, because they have to or as part of a documented process? It's important to keep the speed of decision making as close to where it needs to be done as possible. Are you getting the right information to the right place at the right time and are the recipients empowered to use that information?

Check if you have any citizen roles (roles that aren't in the traditional data teams but are situated within the business – yes, people living in Excel hell working on reports count). It's also useful to know how many people there are as it will help with the justification for some of the changes you may need to make. When you put everyone together it can be shocking how many people are working on producing reports based on poor quality data when they could be doing something that helps the company an awful lot more.

Organisation

Are roles clear and agreed?	Are roles and responsibilities clearly defined?
Do the roles cover all the elements of good data and information management you are focused on?	Does a RACI model exist?
Do they have what they need to succeed?	Are there any gaps which aren't covered?
	Is the RACI model agreed and understood?
	Are the necessary resources available?
	Is it regularly reviewed?

Skills

What type and level of skills already exist for you to work with in relation to where the organisation needs to get to?
Are training programmes in place for both your data and information professionals and for the wider organisation?

This is another area where there is a possibility to merge with another element of the maturity model, like organisation, for example. However, you could be in a situation where you have an abundance of people with the right job titles who all have clearly defined roles and responsibilities, yet none of which match their capability. There are lots of reasons for why this happens but, if this is the case, you would be better off keeping this area separate, at least to start with. You should also keep this element distinct if the company's ambitions differ widely from where they are currently situated. In this circumstance the organisation will have the skills needed to suit where they currently are, not where they want to be.

Your people are a key area when it comes to the success of your data ambitions. We will look at the company culture under behaviour. This section focuses on whether or not they have the right skills and experience to succeed in the journey ahead. Knowing where you are starting from really helps in this area. There are so many ways of covering this gap, from bringing in experts to supplement your teams through to in-house training and development or any combination that works for you. If you currently do everything in Excel then you may have to invest in getting some more data science skills on the payroll. As is the case in other areas, you should not make assumptions here. Just because you currently use Excel doesn't mean that some of your people have used their own initiative and trained themselves in Python. This is, however, exactly the level of enthusiasm that you will need when the going gets tough.

If your HR department has any kind of skills matrix, database, training programmes or record of skills and experience, that is a rich minefield of evidence for this area and worth taking into the workshops with you.

In the workshops themselves you are looking for the general view of the company about where the skills level of people lies. Do you already have the people waiting with bated breath to carry your data forward or do you need to put a full programme in place to make it happen? Is identifying those skills gaps through analysis and providing relevant training programmes already part of a wider initiative or is training not a fundamental part of the offering at your company? This is important to understand when you are looking at how you close the gap. If it has been an ambition of your HR department for years to get something like a skills analysis off the ground but they haven't managed it on their own then they could be a brilliant ally in this area.

Also look at when skills will be needed and who currently delivers them. If you have a really well-skilled organisation that is perfect for what you need going forward but you are overly reliant on external people then you could also be creating a problem for yourself. You don't want to bring in a lot of highly-paid people for something that once it is done it is done and you don't need those skills anymore.

Skills	
What type of skills do you have within your organisation?	Do you understand all the skills needed in order to succeed?
Are they at the right level for your company?	Is a skills matrix in place?
Do you have training in place to upskill or keep skills up to date?	Do you have a skills gap analysis in place?
	Do you have a plan in place to close any gaps?
	Are the workforces engaged?
	Does it take into account future plans?

Metrics

Are you measuring progress and performance measurements, reporting or benefits capture at a corporate level?
Then, are you measuring the right things to drive the kinds of behaviour that you want in your data-valuing culture?

It is a simple truth that people do what gets measured. Measuring something can drive behaviours and sometimes even values within a company so be very careful to only measure things that drive the kind of behaviours you want to elicit from the company. The other side to this is that the human brain can only concentrate on about five things at any one point in time (and usually not that many if one of those things is a source of great focus!), so don't try and measure too many things. Only measure what is important to you otherwise you dilute people's attention and they don't concentrate on the things you really want them to focus on. Link the measurement to what exact behaviours you want to drive. For instance, a measurement like a reduction in how many input errors in a table will (hopefully) drive behaviour around improving data quality but what business process KPI are you trying to impact here? What performance metrics and ultimately what outcome for the business (which should be in the business strategy – this is one way of demonstrating that golden thread that runs all the way from the business strategy right through to the impact you are having across the business) is intended? It is ultimately about results and demonstrating those results?

If you don't understand how to measure something then how do you truly unlock its value? Once you get your head around that then can you check that you are measuring the right things? If you are going to harness all this collective attention on five small areas then you really want to make sure that, firstly, they are the right things to focus on and, secondly, that they are right for the business (in balance with any other metrics that are being measured). The thing to watch out for is whether the metrics are all focused on one area or spread across different areas – a wider coverage. Also check if you are getting tangible measurements – go for pragmatic measurements that can be demonstrated rather than high-level ones that aren't truly measurable. Are the measurements themselves being taken seriously? Do you have any financial measurements based around the data?

You have to demonstrate progress or your buy-in will wilt. Quick wins are great but they only last so long before they aren't quick anymore. Even delivering a series of quick wins will not keep your goodwill going forever. Do the measurements demonstrate clear progress? To do this takes time to set up properly but that is what your quick wins are for – to cover that period when you need time to create your baseline and trajectory.

If the metrics are set up, then are they clear, simple and regularly reported? Do they demonstrate the impact the change in the company is or will be making? Really you are looking for the links between the data measurements and the business outcomes. Making improvements in your data is only useful if it delivers real business benefits, whether they are mitigating risks or creating and enhancing value.

Metrics

Are you measuring your progress and performance?	Are measurements in place to demonstrate progress?
Are you measuring the right things?	Do you track the measurements over time?
Is it driving the right kind of culture change?	Are actions taken as a result of reviewing the measurements?
	Are your measurements spread across areas you want to focus on?
	Are behaviours changing positively as a result of the measurements?

Behaviour

Do you value your data and understand what it can do for you?
One indication of this is how much money you have spent on it and over what time period.
Has it been underinvested or has it been well invested but not succeeded, and why not?

Do words match behaviour, this is about understanding the culture without using the c word. What you are looking for here is the evidence that demonstrates the value the organisation really places on its data. You first need to look on the surface to see what they are openly saying about the value they place on it. Does the company have any communication about data or information, such as one-pagers on how to look after their data?

Are there meetings already set up to look after data with lots of different stakeholders listed as attending? Do they believe they are a data enabled organisation?

Once you have a picture of what the surface looks like you then need to look below the surface and see if it all matches. One way of looking at whether data is already being valued is by following the money. How much has been spent on data over the last one, two or five years? Is it increasing or decreasing? This tells you so much about whether interest is building or waning.

When you find out the spend you also need to get a picture of what the outcomes were. If you discover a situation where a great deal has been invested but no progress has been demonstrated, then there is probably going to be a greater level of resistance to any changes you need to make now.

Other ways of looking for evidence involve seeing if data or information is discussed in any meetings. Are their regular meetings already held to look at data and, if there are, do people actually turn up or do they send delegates?

If there is any conversation about data is it all transmit or a healthier two-way street? Don't let the organisation get sucked into thinking that because there is noise about data that it is being heard, understood and actioned.

Behaviour	
Are the words demonstrated in everyone's behaviour?	How well attended are any data meetings?
Do they act like they understand the value of it?	Do people treat the data as an asset?
	Are data conversations regularly part of non-data meetings?
	Do ideas and issues come from within the organisation without prompting?

Technology

Technology has massive legacy costs associated with it, some of which may not be easily adjusted or when you start digging multiple versions and types of software may be confusing your current workforce.

First and foremost, technology can be a delicate subject to discuss in any workshop. The IT function is a favourite whipping boy of many organisations and the last thing you want to do is alienate them, especially as they will have a part to play in helping move things forward. Technology moves at such a pace and lots of companies are dealing with big legacy issues that they have heavily invested in so can't change course on the head of a pin. It is too easy to blame the IT department if they can't move at the pace needed for an agile data environment. They may well have been dealing with confusing requirements from the business that constantly change and have had to make decisions based on priorities as to what was kept up to date. Don't judge until you have walked in their shoes. The best selling point that you have with the IT department is that you are there to help them. When it comes to all the competing demands from the organisation you will take care of the requirements for any data-related systems and act as the sole focus of corralling lots of different and competing demands. Sell the relationship for what it is, a real win–win.

As well as looking at what the technology estate looks like this area also needs to cover how changes and priorities are made when it comes to that estate. Who makes these decisions and is data a consideration in how new technology investment is decided? Change management is the process for managing change within a project or organisation.

There already may be an investment plan in place for what the company intends to do with technology over the next few years. How does this play into technology that supports a flexible data agenda? What is the overarching technology strategy and how much can be input into it? There is no point in basing your data ambitions on a total and radical overhaul of your technology estate if you are locked into a five-year funding cycle, for example.

This section is about getting that understanding of the technology environment: what applications are helping or hindering your data, what controls are already in place around those systems? It is worth also considering what the business capability is around those systems: what training is in place, is there any support on how to get the best from them not just whether they are working or not?

The focus should be on the business outcomes not the technology, but that doesn't mean you don't have to understand it. At the early stage it is more important that you know what you are working with and where the big gaps are before you make any further investment. For goodness sakes check what you already have access to, you have no idea how many times we went to buy something only to find out that the company already had it but wasn't using it – or something else that does the same sort of thing. You have to be careful with your funding. If you already have something that does 80% of what you want and can therefore spend your funding elsewhere, then you have to decide where you will get the biggest bang for your buck.

There will be significant investment in any IT landscape so pretending you can just ignore it won't work. If you already have a data warehouse or data lake infrastructure, then you aren't going to get rid of that in order to put your favourite tool in place that does nearly the same job. That just isn't practical. If there is a substantial difference then you should look at the longer-term business case.

Is there a technology strategy and are there any established links between any kind of data strategy and the technology strategy? How open is the IT department about accepting your help?

Technology	
Do you have a technical legacy to deal with?	Does your technical estate help your data strategy move forward?
How easy is it to modify your technical estate?	Are good change management practices in place?
What control do you have over this?	Is there a good working relationship with your IT department?
	Do agile working practices encourage innovation?
	Does the data team have the ability to work on innovation projects?
	Does the data team feed into the technical strategy?

Data maturity assessment

Hopefully the chapter so far will help you to understand what you are digging into for each element within the maturity assessment. It really is the understanding of each area that makes the difference, then you can tailor the questions you ask to make sure you dig in to the areas that mean more to your company. Now let's dive into how to run the assessment as this will vary depending on whether it's the first time your company has run this exercise or if it's a follow-up assessment.

If this is the first time you are running the maturity assessment it will be time intensive. There is no point pretending that this is something you can do quickly. If you don't spend a bit of time truly understanding an accurate picture of your data landscape you could well spend your limited time, resources and money working on the stuff that really isn't important. It's not just you that will be really busy during this period, and you will probably have the luxury of this being what you are concentrating on, but as far as the rest of the company goes you will be asking for time and effort from them and will still have their day jobs to do. It might be a good idea to give them a heads up that this is coming, even if it's a vague 'we will need some of your time – about (insert months)', so that it's not such a surprise. Suddenly popping half-day workshops and two-hour interviews into someone's diary is not going to win you any friends.

The other thing to take into consideration when you are planning how to do your data maturity assessment is the number and variety of people you invite and get involved. The more people you can include the better, it all adds to the richness of the results you will get from the maturity model and gives you unbiased results. It also adds credibility to the end result if people feel like their area has been well represented. Just bear in mind that it also covers you for data villains or over optimistic data heroes who can easily swing the results if you don't have a counterbalance to help you get a more realistic result if there are more people from the same area.

Make sure you get a spread of people from across the organisation as well. At a minimum you should have representatives from every department with stronger representation across HR, IT and finance and any heavy data usage areas. If you only add data savvy people, who have their areas sorted, it will appear that the whole organisation is in a great state with nothing to do rather than identifying pockets of good practice in a sea of inadequacy. The counterbalance is also true: if you don't have every area represented in some form then you may miss the pockets of good practice that you can then leverage when you are closing those gaps later.

Another good reason for doing this assessment is that when you are working with that many people, as well as getting information from them, it is a chance to start your education process about what you are trying to do. Never waste any touch point with anyone! Everyone you are working with is a stakeholder in this process, every single one of them is a potential data cheerleader (if they aren't converts already!) so make sure you take the opportunity to evangelise. Don't just book an hour with them and rush through – this is tempting because everyone is so busy– but remember they might not think in the same way that you do. We will virtually guarantee that people will try and cut short these meetings and you will hear every excuse under the sun. Make sure you are prepared for this and have a great story for why you need their time. What are you hoping they will get out of it in the short and long term? Most importantly what is in it for them? It can be the best story ever for the company as a whole but if you want their undivided attention you must have an idea of the art of the possible FOR THEM.

Spend time going through the questions together, slow things down and check that their understanding matches yours. And if it doesn't, why not? Have you missed something? Don't be frightened to course correct as this assessment goes on, remember it is easier to fix things at the start then the day before you deliver one of your major benefits. There are so many times we have had conversations with people and been absolutely convinced we were on the same page when we weren't. Just because we were using the same terms didn't mean we were talking about the same things. It is a common misconception in companies that if you are using the same acronym then you are talking about the same thing.

People who join organisations believe that there is a common language to learn to understand the organisation better. What you are really learning is the nuances of the area within the company you are joining. As we mentioned in Chapter 1, it is like everyone in a country speaking the same language but there being many dialects within it. Everyone is technically speaking the same language but put a Londoner with someone with a broad Scottish accent and they may not be able to understand each other despite both speaking in English.

Ways of conducting information gathering for your first exercise need to be very face to face focused because of the multiple reasons for doing the exercise. From stakeholder management to fact finding, these interactions are valuable. As well as going through the more formal questions, make sure you allocate time to let them talk about what's on their minds – this is brilliant fact finding for later. What keeps them awake at night? What do they think are opportunities? Don't make this stage about data in anyway. Use language that they are comfortable with. You will get to the same end point but they will stay in their comfort zone. Be transparent about why you want the information and what you are going to do with it.

For your major stakeholders a one to one session is best. Give them time (if you can get it) and use it to convince them to be a cheerleader for this initiative going forward. Having senior-level people invested in this will make a massive difference to the support you will get from everyone that sees them as a leader. Visible sponsorship can make or break what you are doing and getting that starts here. Again, remember that while they will be responsible for large parts of the company you still need to focus on what they are interested in, tie in with the 'what's in it for them' idea.

In order to get the best coverage, use workshops to bring groups of people together, try and get them from the same function or with the same focus as they will generate ideas from each other and get a richness that you may miss from just one representative of a particular area. It also helps that you get more validated opinions just in case you get the one person in the department that can't even find the glass never mind viewing it as half empty or the little miss happy.

For the workshops it's important to work with a good facilitator. If you are a good facilitator, then excellent, but if you have more than about five people in your workshop, then take someone else in with you. Having that extra person in there means that they can manage the dynamics in the room while giving you the space to concentrate on what is being said. Don't let anyone blend into the background, every company has its share of different personalities – the reflector needs more space to come forward and the introvert needs a quieter space. Their opinions and views are just as necessary to help you gain an accurate view but if someone isn't concentrating on the personalities in the room you can easily lose these people to the well-meaning extroverts. As already mentioned, that someone should not be you; you need to be focusing on the answers. Are they backed up? Do you need to probe? Did everyone's body language change when someone gave a particular answer indicating that there may be a difference of opinion to uncover? Have you got everything you need? We think you will be busy enough without doing the facilitating and writing up as well.

Whatever the format you are using, the process for each interview or workshop should be the same:

- Explain what you are doing.

- Explain why you are doing it.

- Let them have a short time to air any concerns but keep this limited (otherwise you will have a moaning session on your hands and while this may be great for their mental health it will not be terribly constructive for the exercise at hand).

- Explain what is in it for them. (If you miss any other part of this process, do not miss this one. There has to be something in it for them otherwise you are fighting an uphill battle on engagement.)

- Explain what is in it for the rest of the organisation (the higher up the organisation the people you are working with the more important this should be to demonstrate).

- Show a strawman of what you are trying to achieve. Give an idea of the chart and the diagrams you are going to use. Do not under any circumstances be tempted to use examples from previous workshops as this can bias the people in the room – favourably if they like the people or department and want to agree with everything on the chart, or vice versa. In either circumstance you are not getting unbiased representation.

- Go through specific questions based on the criteria listed above making sure you have a few questions for each area (just one or two will generate skewed results). Try to get to a consensus score if you are in a workshop but make a note if this is impossible. There is value in the disparity that you need to investigate when you start using the output of the maturity model. Yes, it means you need to take it into account when you are writing up the model, but it gives insight into possible data heroes at work in the company already or delusions of data grandeur that you need to correct.

- Have an open discussion time so they can say what is on their minds. Always ask the question 'anything else you want to add?' This is especially important for the reflectors as they might have a number of points they have been thinking about.

- Go back over the output you think you have from the workshop just to confirm they are happy with what has been covered so far.

- Remember to say thank you, politeness should not be overrated.

As you go through each question, try to gather as much evidence to back up the score that everyone is agreeing on, whether that is finding out where something exists (so you can collect it and make sure you really have what you are supposed to have) or that the lack of it existing points to a certain score. If an artefact exists, then you will need it in some way when you start to put changes in place so there is no wasted effort in collecting it all now.

Some areas are subjective so you won't be able to gather tangible evidence. However, the more you can gather the stronger your arguments will be going forward. It's hard to argue with cold hard data but it's also very hard to argue with a compelling story. Don't treat the subjective parts as second class, they will help you understand what is critical to changing the behaviour and creating those data cheerleaders you need. Make sure to collect why the subjective parts have been decided upon – this will help you when you have to justify it later.

Vary closed and open questions and if it is an area of concern don't worry about asking both types of questions – for example, 'Do you have a data strategy?' leading onto 'Is it any good?' – in order to clarify your view. This is why you need to have an element of flexibility built into the model so you can probe into the answers rather than follow a rigid set up that will only constrain you.

Use the workshops and face to face time as much as you can the first time you run this maturity model but don't rule out utilising questionnaires if it is the only way of getting information out of someone. Something is better than nothing and, if you have tried for the face to face and given it your best shot, rather than damaging a possible business relationship by moving into nagging territory just take a step back and accept what you can get.

Eventually you end up with something like Figure 2.1. Now if you have been really clever you will have constructed your workshops so that you can have a chart like this for each of your organisation's departments, which you then pull together for the maturity assessment score for the whole company. The point of this is so that you can (subtly) pit each department against each other. Senior people tend to be very competitive animals. They don't really like it if you have a chart that clearly demonstrates they are dramatically worse at something than a rival who runs another part of the company. They do, however, really like it if they are the one that gets to lord it over the rest of their colleagues and demonstrate why they are so good at implementing a framework, for instance. Think of it as introducing a competitive gamification element into who can improve their area the fastest. And in the end, everyone wins, what could be better than that?

What do you do with your maturity assessment?

Having a maturity assessment is a great start but what do you do with this data or at least how do you represent it in a way that is useful to you in your storytelling journey as a data leader?

Effectively what you end up with at the end of the assessment is a table full of data, qualitative and quantitative evidence to back up why you have those numbers and some stories and examples from different parts of the business. This is leaving to one side all the other very useful things you have from the maturity assessment process such as building or strengthening relationships with your stakeholders.

There are obviously different ways of representing this data and you can use whatever makes sense to you. The one we prefer is the radar diagram, if you take the table illustrated in Figure 2.1 you can turn it into the diagram in Figure 2.3.

It is a really simple way of presenting this information in an assessable way. It only, however, gives you part of the picture, the other important part to figure out is your organisation's ambition when it comes to data coupled with an idea of their current and near future capability.

Breaking this down into ambition first, on a scale of 0 to 5, your first instinct when you start looking at this is to want to be the best – jump

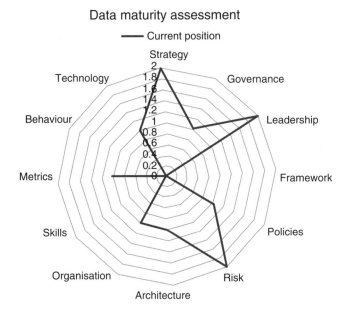

FIGURE 2.1 Example of a radar diagram from maturity assessment

to 5 and be an exemplar in data that other organisations look up to. The really honest truth is that this scale is a bit like that popular logic puzzle: if a frog starts on one side of the pond and every jump covers half the distance to the other side of the pond how many jumps will it take to the get to the other side? The answer is never, but the frog will make increasing smaller jumps, expending effort every time with increasingly smaller results. Striving for 5 might seem like a noble aim but do you really want to throw every resource you have into reaching a goal that, when you get there, can probably be bettered?

That's not really the point of the ambition side of the scale – it's to understand the relative importance to your organisation of each point of the maturity assessment in relation to each other. Where do you want to expend your resources and effort? We haven't yet found anywhere that will give you everything you want when it comes to resources, money and focus for making changes. And if there was then we wouldn't have asked for enough. There will be compromises and priorities to consider. This gives you an idea of which areas need your full attention and which ones can wait just a little longer.

It is possible to talk to each workshop about the level of ambition for each point to help you get a collective picture. This will take time from the discussions about where the company currently is and won't necessarily add much value. The factor that you should be clear about, that the people in the workshop may not, is the organisation's current capability to address the gaps. The company could think that getting to a 4 on the metric scale will really drive them forward but, if the whole organisation doesn't already buy into working with metrics as a matter of course, you might be better off aiming for a 2 at the beginning and focusing on getting your framework and policies set up to a level 3 before you reassess where you are. You aren't setting your ambition at a level 2 forever, just as a starting point. Consider it your interim ambition position based on knowledge of the current organisation as a whole and their ability to change. Remember you can change this but it helps to focus people's minds on the direction you are taking the business.

Overlaying your ambition against the current position will get you something like Figure 2.2. Use the comments you picked up from your workshops overlaid on this as well to start to bring this to life for everyone.

Figure 2.2 clearly demonstrates the gaps between the current situation and the ambition of the company and, as you can see, there will probably be some pretty big gaps. It's just doubtful that you can fill all of these gaps at the same time so you will have to prioritise the order you tackle them in. We will go through how to do this later.

Figure 2.3 is a more advanced example of what you can end up with to demonstrate your overarching output from the maturity assessment. This visual representation lets you clearly demonstrate the radar diagram

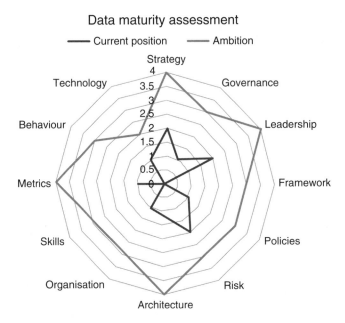

FIGURE 2.2 Ambition against current position example

alongside the key points you want to draw from the diagram. It is a great tool to start the storytelling with your stakeholders.

Slight aside

Since we aren't fans of wasted effort its worth mentioning that this baseline maturity assessment will also be used as part of your governance and assurance activities later. Any throw away spend needs to be kept to a minimum! However, when you do repeat the exercise it isn't as intensive or time consuming as this first time – hence why we keep talking about the first time you do it!

Having this as an annual exercise is a great idea because you can then use the radar diagram to demonstrate your progress. However, if you try and repeat the exercise exactly and use the amount of time you did the first time round you might have a rebellion on your hands. With this pass

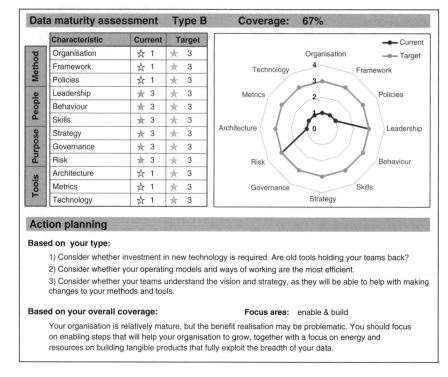

FIGURE 2.3 Representation of results from a data maturity assessment

through, for the people who have done it before, it may be easier to ask them to complete this themselves, either as an in-department workshop or as a questionnaire. It depends on the level that you are comfortable with. Using a blend of lots of different methods here gives you the flexibility to still engage with a good breadth across the business without it involving quite as much time and effort.

With anyone who hasn't been part of the initial maturity assessment we would still recommend you spending the time with them to help them understand the wonderful world of data.

Now this is where the fun really starts…

Chapter 3

Making the Change Happen

This stage in the transformation process is all about filling in the gaps, bearing in mind all the different parts we talked about in Chapter 2. It's great that you now know where you are and where you want to be but this chapter deals with the practicalities of how to improve your state. So, it is about knowing where you are going and how you intend to get there. The word 'intend' is really important here. It's great to have a direction, and even an optimum road map, but it doesn't account for any unexpected bumps in the road or shortcuts that you can only see when you happen upon them. We don't want you to get so tied up in following a process that you don't use your own intelligence and insight in order to know when to adjust course to accept a fairer wind. This stage helps you to understand the plans and strategies that you need to put in place to set up for success. But don't think that you have to follow them rigidly! If you have a great idea that wakes you up in the night after the plan has been agreed – don't forget about it, use it. At the very least address the idea and see whether the change is worth any disruption it might cause. Ensure that you are making your decisions consciously and not blindly following a plan that was probably a great idea when you came up with it but may not take into account each small deviation from the path.

Remember, proper planning prevents poor performance. Understanding the context is great, you have a firm position to start from. You are standing in the shopping centre staring at the great big map, usually with people looking over your shoulder so they can see the map as well, and you have a great big sign in front of you telling you exactly where you are. You've done some great work to this point, you even know which shop to go to first (i.e., which element is your priority). There is a big difference between knowing where you are and knowing where you want to go. Understanding you have a problem is one thing but this stage in the process is about what you are going to do about it.

It used to be that a larger company had an advantage over a smaller company as it had volumes of data that provided an ability to make higher quality decisions. However, that trend seems to be less clear cut than it used to be. There is an issue now where the volume of data has increased so much that it's like an avalanche causing people to make decisions in spite of the data rather than because of it. Smaller organisations or those set up to take advantage of the data now have the upper hand. Going through this process helps you take advantage of your data and gets it working for you again, rather than feeling like you are drowning in a sea of digits. The focus has to be on driving an improvement in the quality of decisions. Every small decision has a cumulative effect.

One important thing to note is that nothing progresses at exactly the same pace. Obviously you will have a critical path, so some things have to happen before others – for example, you have to put a key in the lock before you turn it – but not everything needs to be so linear. Identify the critical path items that must happen in a specific order and then look at what else can take place in parallel. In most cases you don't need to wait for one thing to finish before you start another one, multiple resources or multi-tasking can accommodate more than just linear thinking.

We talk about this being a programme but you need to set expectations carefully as the programme will create the change but the organisation will need to adopt that change. A new target operating model around governance, for instance, will be added or modified and there will be ongoing activities to ensure that this is not allowed to fade away. So, we will cover strategy in its own right but, considering the idea that you will be making a transformational change to the organisation, you need to think carefully about what you are trying to achieve.

Starting with a programme mandate helps define what it is you are trying to achieve and gives you something to start with and talk to the organisation about. Since you are beginning with the programme, you might as well start right and define the change you are trying to achieve. Essentially, a mandate needs to cover the basic definition of the aims and, very importantly, the benefits that you are looking to achieve. Having a clear articulation of the expected and target outcomes and an understanding of the various impacts such as constraints takes time but will help you steer a steady course. Think about it in terms of slowing down in order to speed up. A good start means you will have a much easier time on your journey going forward. Think about your mandate in terms of the shining light in the dark to help you head in the right direction. It's not the detailed strategy that you need to see the path but it is a great catalyst to get people on board with your vision of changes. It also helps when setting expectations about what the company is signing itself up for.

The note of caution here is that if the company doesn't believe up front that the programme will be delivered, everything will appear to be fixed and everyone can go home. Part of the setting expectations exercise needs to be about the understanding that, while you are starting from a programme point of view, you are creating a long-term change – so there will be a business as usual function that will either be created or modified going forward. The last thing you need is to have the business thinking that if it invests attention now it can abdicate its responsibility again in the future.

Methods and methodologies

There are lots of different methods and methodologies that can help you elicit a change within a company. Like with technology, we are very agnostic about which one you need to use. We are big advocates for using the right horse for the right course. That said, while we aren't proposing that the agile method of working is the only one you should use, it is important that you work in an agile fashion. For anyone who understands the full agile method it uses incremental, iterative work sequences that we normally call sprints. You don't need to follow this slavishly as a method for what you are doing, rather focus on thinking in an agile fashion. What can you get from thinking incrementally and iteratively? However, you form the workstreams you are using for your transformation, you don't need to set it in stone before you start work on the other workstreams.

Think agilely – you might not need to throw yourself into a full-on Kanban type of activity that you pour over (although we do like this particular way of charting progress) but be agile in thought. As long as you embrace some of the key areas around working flexibly and not rigidly you can benefit from this way of working.

Based on the work you have covered in Chapter 2 you have an idea of the gaps that you need to address but now you need to prioritise those. Don't try and fill all the gaps at the same time. Look at the urgency of any gaps, are there any areas that must be solved now or else give you a very difficult time? Are there any areas that, while they are a big gap, can wait until you get some of your foundational elements in place?

You can't possibly do everything at once. For one thing, think about your capacity and that of your team. There is only so much that people can cope with, especially if you want their best work. When you are doing any kind of transformation then you need people to bring their A game. You and your whole team will be the face of the transformation, and having real people associated with the transformation will help when times get tough.

You also need to understand the pace of change that the organisation can cope with. Are they already coping with change fatigue or ready and raring to go? There will be a heartbeat to the company, a way that things work and a pace that they either like or can cope with. While you can speed things up in some cases, trying to constantly fight against the natural tempo of an organisation can be unbelievably exhausting both for you and them. Choose your battles, there will be plenty of things that you really need to expend your energy on.

From the maturity assessment model we categorise each area into a quadrant: purpose (always start with purpose), people, method and tools (see Figure 3.1). Alongside your priorities, among which will be a million things that you are dying to get started on, try and get improvement actions within each of the quadrants in order to make sure you have

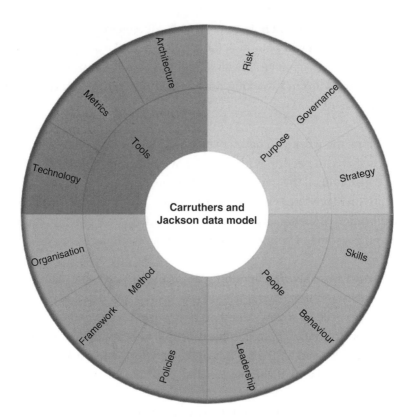

FIGURE 3.1 Carruthers and Jackson data model

a balanced change. Too much of any one of these areas means you are loading your dice too heavily and your overall results will be skewed in one direction. For instance, if all you worry about is the policy and the tools, there aren't enough people elements to have any impact on your hearts and minds battle so you will fail to engage. Just looking at purpose doesn't put the actual changes in place to have the impact of your purpose, just like having a strategy without any actions doesn't make anything happen.

All of these things need to be considered when you look at which areas to target first. And, while we always say that it can be different for each company, one big rule to follow here is that strategy must be one of the earliest (if not the first) thing considered. It doesn't have to be complete before anything else starts, it's doubtful that you will have the luxury of that much time, but the work has to start as it underpins everything else. Remember, get your basics right first.

In order to make it easy we've broken down each of these areas to take you step by step through how to improve in each area.

Chapter 4

Purpose

E verything starts with a purpose otherwise why would you bother doing it? Remember everything we said about the data underpinning the business strategy? Well that is what this stage of the transformation process is ensuring you do. Not only that, it also helps frame the direction for all your other data activities so make sure that you pay attention to this area and incorporate it as a fundamental part of your data journey. By looking at strategy, governance and risk you are definitely getting a solid foundation for the transformation you are taking your company on.

Strategy

Lauren Walker, CDO at Dentsu Aegis Network

Getting your data strategy right is always a matter of perspective. In my 15 years in the data industry, I have worn various lenses: the data aggregator/value added re-seller lens, the tech vendor lens and now the media/marketing domain lens. As a chief data officer, I am thrilled to see every shape and size of organisation believing in the mantra, 'data is transformational'. My clients and my partners are investing in data-savvy employees, partnering with consultancies, with start-ups and with enterprise tech vendors to modernise, automate and innovate.

From my early experiences at ChoicePoint (now part of Reed Elsevier LexisNexis) and through those formative ten years at IBM, across information management, Hadoop, analytics, Watson and the Weather Company, I've enjoyed creating, executing and advising on a variety of data strategies. In my data-loving eyes, the right data strategy has always had the power to transform, but 20 years ago (1) not as many organizations knew how to do it and (2) the technology and techniques we have today were not widely available.

First things first, before you even think about putting a data strategy in place (and, make no mistake, you need a data strategy) make sure you know what the business strategy is. What directions is the organisation heading in? Who are your main competitors? What are your current business trends? What are you currently doing and what do you want to change? This is where the key to data enabled rather than data driven comes into play. If the organisation is a start-up then data driven may be the right term for you to use. However, in the majority of cases you are trying to change an organisation that is already in a state of business as usual so enabled is the better word to use. Your data underpins the company, and the company already exists so what do you want to do different, how is your data going to enable this change?

You must link to your business strategy. If your organisation doesn't have one then it is down to you as a leader to instigate having one. You should help to write it, encourage others to get on board with what your company is striving to be. You can have the best data strategy in the world but if you don't know what your company is trying to achieve then you will still end up going around in circles chasing your tail.

One of the comparisons that often pops up when putting in place a data strategy and making changes is that it's a bit like trying to perform open heart surgery on a marathon runner who still has to complete the race. In fact, what you are really trying to do is look at how you can help the runner but still make sure they cross the finish line – like providing regular water stops and the right kind of snacks to keep them focused.

As much as we (data professionals) like to think that data is the start and end of the universe it may come as a shock to learn that not everyone thinks like that. (Yes, we know, we find it strange too.) We have to recognise that business actually existed before the focus moved to data and organisations became data enabled – in some cases they've done really well. What you are doing when you put in place a data strategy is looking at how you underpin the business strategy. We don't want to create a data strategy just for the sake of having one. Data in an organisation doesn't exist just to exist, it has to underpin what the business is actually trying to do. As Lauren says, data has the power to transform, we just have to make sure it is in the right way. Keep the phrase 'what's in it for the business?' in your head when you are writing the strategy. We use different variants on that phrase all the way through the book but they all essentially boil down to 'what's in it for them?'

Understanding the context of the business has already been covered in Chapter 3 and that will really help when you are developing your data strategy as you should have a great picture of where you are and an under-standing of the ambition in each area. A data strategy is one of those inte-gral things that has to be part of your plans. It may not be one of your overall top priorities (and if it isn't – why not?) but it is really important that it is something you do put in place.

A central pillar to direct and support the data-driven business transfor-mation is a data strategy. This is a document (we will discuss later in this chapter whether it is a document or something else) of what format the data strategy will take, that sets out the vision of the end state of the data transformation and provides the route map to get there.

When we have discussed the data strategy previously we were always very clear that it must align with the business strategy, or the desired business outcomes or goals as expressed by the owners, board or executive committee. In the context of a data-driven business transformation there should be a document, or artefact of some kind, that at a business level describes and defines the business transformation that will take place, how it will be driven by data and what the outcomes or the transformed business will look like. In detail this should address the three core elements of people, processes and technology, looking at all operational functions, support functions, service functions, back-office functions, supply chain, sales, marketing, compliance and risk.

Many transformation programmes fail because the highest-level business 'vision' or strategy document is not worked up in enough detail, with elements of it delegated to separate business units or services. In this case the business strategy for transformation becomes siloed. It may be necessary to develop these through separate departments to draw out the detail, but they have to be brought back together at an enterprise level to set the enterprise strategy, to bring cohesion and align them. Once this has been completed a data strategy can be developed with the focus of driving and delivering the outcomes or the vision of the business strategy.

A data-driven business transformation will not be delivered by simply writing a data strategy, but the gradual delivery of this strategy will inform and nurture the business strategy. The concept of dynamic data-driven transformation depends upon this. As the data strategy is delivered over time, more insights will gained from the data. Or the way that data is collected, managed and exploited will evolve as the data strategy is rolled out, and this in itself should drive the evolution of the business strategy.

In reality, and in the most innovative and disruptive environments, the initial business and data strategies may be written side by side, each informing the other. On the business side what is desired and on the data

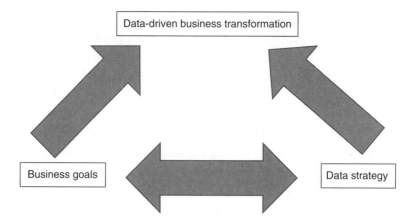

FIGURE 4.1 Alignment of data-driven business transformation

side the art of the possible. As Figure 4.1 demonstrates, the business goals and the data strategy give you the direction for your data-driven business transformation.

It is essential, in our view, for any business to have a data strategy, even if it is not embarking on a data-driven business transformation. Which organisation does not have a financial strategy? Or an HR strategy? Or an IT strategy? These each address a key asset of the business: money, people and technology. Data is also a core asset for a business, therefore a data strategy is essential. A data strategy will deliver at the very least efficiencies and a more effective capability. It will enable cost cutting and in the best cases deliver return on investment (ROI). In some cases, the data strategy may even be cost neutral – funding itself by delivering efficiencies – and, in other circumstances, it may even be funding positive – where the cost reductions delivered by the data strategy outweigh the costs of delivery of the strategy.

We have written, in *The Chief Data Officer's Playbook*, and spoken about data strategies at some length previously. We have discussed the importance of the data strategy for the CDO as an essential tool and guide to avoid the hype cycle. We have also discussed the data strategy in the context of the first 100 days as a CDO and the next 300 days as a CDO. These are important lenses to use when developing and delivering the data strategy.

We don't want to repeat what we have written already, rather we want to take the concept of a data strategy forward, evolve the thinking and understanding and, at the same time, place the data strategy in the context of delivering the data-driven transformation of an organisation. Many sources, us included, talk about data strategy in terms of how to create one and how to deliver it, but we wish to explore how the data strategy can and does enable the data-driven transformation.

Data strategy tracks

One of the very next things to understand when talking about strategy is that you don't have to fix it in stone right at the very beginning, in fact we recommend that you don't. Don't assume you know everything from the start. Just because you are pointing the company in the right direction doesn't mean you won't need to refine the direction it and its people are heading in. From the mandate you will look towards an urgent data strategy next. A programme mandate not only gives you a really high-level understanding of where you are taking the company but also a headline for the urgent or immediate data strategy.

Previously we have suggested that a data strategy could be made up of two concurrent tracks; the immediate data strategy (IDS) and the target data strategy (TDS). The IDS is to address the immediate data issues of an organisation, the burning issues; it delivers some early wins for business value and it is needed to demonstrate the art of the possible in order to educate the organisation. It is also important in the IDS to start the cultural change towards being data driven and data centric. The TDS is intended to deliver a more strategic and complex end state that is only achievable over a longer period of time.

We have now re-cast these two concurrent tracks into three by adding an urgent data strategy (UDS) in front of the IDS. This evolution of the concept of the data strategy into three tracks running over different time periods now takes into account a more complex and dynamic end state within the TDS, which should embrace and embed dynamic data-driven transformation (D3 – discussed in detail in Chapter 10) rather than delivering a fixed end state vision (see Figure 4.2).

Even though these three strategy tracks are concurrent, the level of activity in each will change over time. It is inevitable that as the activity level changes in each of the skill sets required will change as well. And, again, people, process and technology will play out across the evolving strategies. A constant communications plan and narrative is also required to enable the organisation to comprehend where they are in the processes, and to understand any changes in the shape of the graph. It is inevitable that no two organisations will have the same graph, as presented in Figure 4.2, as a generic interpretation. However, it is essential that the evolutions of the UDS, IDS and TDS are mapped out over time and their interrelationship understood. One interesting feature of the TDS as it reaches maturity is the plateau to the right of the graph, this represents the D3 state.

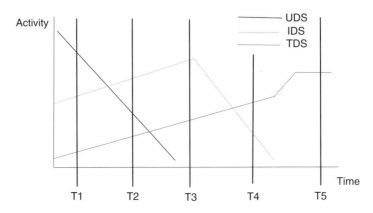

FIGURE 4.2 Influence of each data strategy

The exact nature of an organisation's data strategy will be driven out of the findings of the data maturity assessment and the desired shape that an organisation wishes or perceives that it needs to achieve to deliver a data-driven business transformation.

It is helpful to understand some of the characteristics of the stages of the evolution:

T1 – At this point the data strategy is early in its delivery. The CDO, or the data leader, may be in their first 100 days. Most of the activity is on the UDS, which will be addressing the urgent data issues within an organisation. These may be: some initial data governance, data lineage, reducing the over dependence on spreadsheets and shadow IT, tackling the high-profile burning issues. The features of the UDS by nature will be tactical, and there may be some regret spend or time to solve or remediate burning issues, at this point you may be building up technical or data debt, which then has to appear on a backlog to be addressed in either the IDS or the TDS. At T1 there is some work laying the foundations for the IDS, which might be laying the foundations or some level of automation in data processing and analytics. It might be the early stages of developing an organisational design for the data and analytics function. The activity at this stage, which is focused on the TDS, will be mostly around communications and narrative, working with stake holders to create and narrate the end state vision. T1 should demonstrate high levels of engagement from most parts of the organisation – people will probably be relieved that 'data is going to be sorted out and taken seriously'. At this stage employees will not feel threatened by the data transformation or the changes it will bring.

T2 – At this stage the UDS is winding down in activity and the focus will be shifting to the more structured and strategic IDS. Data governance and management should be becoming more mature and a data culture should be building across the business. The IDS should be adding significant business value. The early foundation work of the TDS should be underway, this will be laying down the essential building blocks for the future to deliver and sustain the data-driven business transformation. This may be addressing new methods of data integration, data transformations, data storage and analytics. The TDS at this stage might be in a very innovative stage, trying many things and failing fast under the supporting cover of the IDS. T2 is an important stage for the business to learn data-driven approaches and to buy into the concepts of agile deliveries and new styles of working. This stage will be exciting and fast paced with activity on all three tracks simultaneously. To sustain this high level and variety of activity there may be the need to rely heavily on suppliers, contractors and/or consultancies. One key tip at this stage is to avoid as much regret spend or time as possible in the IDS and don't build up too much technical or data debt.

T3 – By T3 the UDS activity should have been completed and the data environment within the organisation stabilised. The IDS will be hitting peak levels of activity with the activity on the TDS increasing and becoming less focused on proof or concepts and more focused on minimal viable products. This is the point at which the data strategy as a whole may come under most threat from across the organisation. At this stage it will be 'disruptive'. It will be changing the processes of the business, how decisions are made and the way in which people work. Skill sets within the organisation will also be changing. At this point the data strategy will be reaching maximum, or certainly higher, levels of budgetary or forecast spend. The TDS will also probably have failed in a few places as part of the 'try a lot, fail fast' concept and the 'cynics' in the organisation will focus on the failures. To address this the CDO must be strong on the narrative at this point and clearly communicate the successes and the vision.

T4 – At this crucial stage the data strategy should be starting to deliver significant data-driven business transformation. This is the point at which the end vision of the TDS should be starting to evolve and pivot in alignment with the changing business goals. This stage may require the greatest skill in business change management as the organisation reshapes around the data.

T5 – This stage is peak business transformation as the TDS begins to deliver maximum and significant outcomes. It is essential during this phase that the organisation hits D3 to sustain the data-driven transformation. By this stage the data culture should be ingrained into business and be at the heart of decision making and operations (see Figure 4.3).

The rate of data-driven transformation will evolve across T1 to T5, with maximum transformation not being reach until the TDS is the main focus of activity.

In terms of artefact, we have discussed the data strategy being a 'document'. The actual form of the data strategy isn't really very important but there are a number of key features that should be achieved and these might determine the best route to publish and share the data strategy.

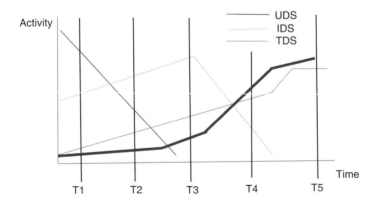

FIGURE 4.3 Indication of where the data-driven transformation takes place

1. The data strategy should be owned by the CDO on behalf of the whole business.

2. It should be endorsed and supported by the board, executive team or CEO, whichever is appropriate, to ensure that it has authority and significance across the organisation.

3. It has to be published and shared.

4. It should be in a form, or multiple forms, and language that make it accessible and understandable to the whole business.

5. It must be an organic document that is capable of being flexible over time and responds to changing business needs and expectations.

6. It should evidence alignment with the business goals and be clear about the business transformation that it will deliver.

7. It has to address people, processes and technology.

8. It must describe the end state of dynamic data-driven transformation.

9. Technical documents should be placed into 'appendices' so that the main part of the document is readable and understandable as a standalone artefact.

How does the data strategy fit into the context or objective of data-driven business transformation? For a while many organisations and data professionals have almost viewed a data strategy as an end in itself, and that 'goodness' for the business will flow from it. At the very best practitioners have sought alignment with the business goals but have only seen a 'passive' relationship to the business goals rather than an 'active' and 'proactive' influence to drive them. To enable a data-driven business transformation the data strategy should be front and centre. It must not be written or framed in a passive manner, it should be direct and proactive to drive the transformation.

The graph in Figure 4.3 shows the link between the data strategy and business transformation. As the data strategy passes over time from T1 to T5 business transformation increases in pace and significance. It is the insights in the data, the automation in the data processes, the growing data capabilities that are developing as the IDS and TDS in particular are being delivered across T1 to T5 that are enabling and driving the business transformation. The concept of capability is very important. The TDS in particular should be growing the data capability. This is everything from the skills in the data and analytics team of data engineering, data architecture and data analytics, to the increasing capability of data science and advanced analytics. It is also the growing capability of data technology within the organisation. It is important that the IDS and TDS deliver increasing data technology capability and unleash growing potential and insight in the data. The growing capability also has to be in the data literacy across the organisation. It is these growing data capabilities delivered through the data strategy that drive the business transformation.

Therefore, the delivery of the data strategy is the enabler for the data-driven business transformation. However, it would be wise to learn from the past. As suggested by Tom Graves from Tetradian Consulting, we have been here before.

Digital transformation?

We've been here before.

And if we're not careful about it, as enterprise-architects and others, we risk making an even worse hash of it than we did on those previous times.

Oops…

But what is 'digital transformation'? There are so many arguments on this that…

(continued)

Yeah, this one's a bit difficult – What we can say is that it's about some kind of transformation, probably in some kind of business-context, that involves something that some people call 'digital'. Would that do for now?

Or, to put it another way, it's just another instance of what happens when some kind of new technology opens up new opportunities to do some aspects of some story – usually a business-story – in a new and different way.

Which is a great idea. Usually. At the start, anyway.

But there are at least four ways to get it badly wrong:

- We get over-excited about the technology – forgetting that a business-story is always, first and foremost, about people.

- We try to make it benefit only some players, at the expense of others – forgetting that a business-story only works as a whole.

- We fall for the delusion that the technology can do everything – which it can't.

- We start to believe that the technology will give us control at last, over everything – forgetting that, in the real world, 'control' can never be more than a comforting delusion.

And unfortunately, any or all of those are exactly what happen, all too often…

To illustrate, let's take a simple model of the relationship between the organisation and the broader shared-enterprise – enterprise as 'bold endeavour' or shared-story – within which it operates [see Figure 4.4]

As a quick summary:

- *Inside-in* is the internal workings of the organisation – a 'black-box', to anyone outside

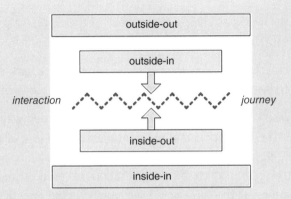

FIGURE 4.4 A simple model of organisation and enterprise

- *Inside-out* is how the organisation presents itself and its offerings to the outside world

- *Outside-in* is how the outside world sees the organisation, and makes its requests of the organisation

- *Outside-out* is the overall enterprise-story, and all the stakeholders of that story, within the further context of the wider world

At the **inside-in** level, probably the classic early example of 'digital transformation' is **business-process reengineering** – the great 'deus ex machina' of the early to mid-1990s. It was a great idea: use new technology to take over the tedium of back-office work, and free people up to do more of the work that machines could not do. And there were ways to do it right. But for most, it went wrong, very quickly:

- It was hugely over-hyped by IT-system vendors and the big-consultancies.

- It was sold as an easy way to cut costs – particularly employee-costs.

(continued)

- It was hyped as being able to do all of the work, via predefined 'executable business processes'.

- It was sold as the way to control everything in the back-office – to force everything to conform to simple lists of 'business-rules'.

Otherwise known, in the real world, as a guaranteed way to screw things up. The IT-vendors and big-consultancies made a killing again, of course: but for almost everyone else, it was a hugely expensive mistake – and one from which some organisations never recovered…again. Not to mention all the careers and lives ruined in the orgy of 'cost cutting' in which so many of those organisations had indulged…

The crucial point was that the focus should never have been on the technology alone – instead, much more on the actual tasks to be done, as a whole; and even more, on the people, and the overall people-story. And yet, in most cases, that last point was ignored or glossed over at the time, as Michael Hammer later ruefully reflected:

'I wasn't smart enough about that. I was reflecting my engineering background and was insufficient appreciative of the human dimension. I've learned that's critical.'

A lesson learned? – we would hope so…

Except that, a decade later, we saw exactly the same happening again, this time at the **inside-out** level, with the rise of **web-technologies** – another great 'deus ex machina'. It was a great idea: use new technology, together with those business-process technologies, to help organisations open up a new kind of conversation with the wider world. And there were ways to do it right. But for most, it went wrong, very quickly:

- It was hugely over-hyped by IT-system vendors and the big-consultancies.

- It was sold as an easy way to sell new stuff to new markets – a new marketing-channel via which to trick customers to 'buy products and crap cash'.

- It was hyped as being able to do all of the work – 'build it and they will come'.

- It was sold as the way to control the market – to 'possess eyeballs' and suchlike ideas.

Otherwise known, in the real world, as a guaranteed way to screw things up. The IT-vendors and big-consultancies made a killing again, of course: but for almost everyone else, it was a hugely expensive mistake – and one from which some organisations never recovered… again. Not to mention gaining the ire of entire markets in the process…

The crucial point was that the focus should never have been on the technology alone – instead, much more on the actual tasks to be done, as a whole; and even more, on the people, and the overall people-story. The catch, perhaps, was that it was a different people-story than one that most large organisations were willing or able to understand, as Doc Searls and others explained in the Cluetrain Manifesto:

Most corporations … only know how to talk in the soothing, humorless monotone of the mission statement, marketing brochure, and your-call-is-important-to-us busy signal. Same old tone, same old lies. No wonder networked markets have no respect for companies unable or unwilling to speak as they do.

But learning to speak in a human voice is not some trick, nor will corporations convince us they are human with lip service about 'listening to customers.' They will only sound human when they empower real human beings to speak on their behalf.

A lesson learned again? – we would hope so…

(continued)

Except that, a decade or so later, we're seeing exactly the same happening again, this time at the **outside-in** level, with the rise of so-called **digital transformation** – another great 'deus ex machina'. It's a great idea: use new technology, together with those web-technologies and business-process technologies, to enable citizens, customers and others to build their own conversations with the organisation. And there are ways to do it right, which include:

- We need to mitigate against the hype about the technology itself – perhaps particularly from the IT-vendors and big-consultancies...

- We need to acknowledge Conway's Law, and recognise that for digital transformation to succeed, we need to change the way the organisation communicates with itself.

- We need to be aware that not everything can be done by IT – and design parallel processes accordingly.

- We need to acknowledge that, whatever technology we use, it's always more complex than it looks – and that we need to keep people, not technology, at the centre of every would-be 'solution'.

Yes, in each case, the technology is important, as an enabler: the new options probably wouldn't exist without that new technology. In that sense, to quote Andrew McAfee, 'it's not not about the technology'.

Yet it's also not about the technology itself – and that point is crucially important. Instead, **so-called 'digital transformation' succeeds only when it's mostly about people and their needs**. We forget that fact at our peril...

Tom Graves (20 January 2016)

To bring this into context on data-driven business transformation it certainly isn't about the technology, we have ticked that one off already, but neither is it all about the data. The data sits in the context of people, technology and processes, but we would propose that for the foreseeable future data will be the driver. We must, however, avoid the mistakes of the past and NOT focus solely on the data to drive the business transformations, it requires the support and enablers of technology, people and processes.

Going into each of these areas in more detail and taking you step by step through building up what you need to start will help break this process down for you.

The UDS is a reaction to what you face when you walk in the door and as such may well be one of the key drivers for your change happening in the first place. Has a catalyst happened that you have had to respond to and, if so, what was your grand plan?

The IDS should be your tactical response to our direction of travel. It provides support for your immediate needs, lays your ground work for the next steps and pulls together what resources you can currently get.

In your IDS you need to address:

- Stabilisation and rationalisation of the existing data environment

- Data culture and governance

- Existing and immediate data and IT development initiatives

- Data exploitation and integration

- Data performance, quality, integrity, assurance and provenance

- Data security

Pull together your data heroes and understand your data villains. If you look at what your data heroes are doing that can really lead them to form part of your quick wins or, as we actually prefer, minimum viable products so you can minimise any regret spend. Don't do anything that is wasteful if you don't have to. This is all about getting the best from your immediate situation. How do you hit the ground running? There are lots of good bits of your organisation that you can use, so don't 'throw the baby out with the bath water'. It won't be optimal but you might reach good enough for now. We often use the phrase 'perfection is a disease' because we get so caught up in striving for everything to be absolutely perfect that we forget that things still have to work in the meantime. The world cannot stop turning while you figure everything out and then start again once you are ready. It would be great but, while you were off figuring it all out, if everything stopped you wouldn't have a business to come back to. If things can work and you are able to get some value out of them why wouldn't you? Sometimes this can be achieved by simply reorganising things or getting the right people involved. Think about it in terms of milking the situation you are currently in.

Where you are dealing with the IDS, you are pulling on all the current extreme levels of complexity, distribution, diversity and scale that you already have to deal with and then thinking about how to integrate and maximise the value we get from this – milking what you already have.

In a target data strategy you are looking much further into the future. This is your aspirational future state that you are conveying, which will help you win the hearts and minds battle – it gives you your battle cry. This strategy is more about evoking an emotion, it fuels your data story and gives you the words to use when you are looking to inspire people.

Again, try not to over engineer what you put into either strategy, careful consideration of a few words can be much more powerful that creating a novel.

How do you actually go about pulling your strategy together? We have a basic process that we go through to consolidate all the different ideas, areas of focus and ambition for your company that we will take you through step by step. This is a really simple four-step process that we use when we pull together a data strategy for a company. We don't cover how to do an UDS here as that is a by-product of something that has happened – a reaction – so you will either already have one or will not need one. We will guide you through the immediate and target data strategies.

Initially you need to understand your position, which you will already have done when you did your data maturity assessment. We also use the same elements that we used within the maturity assessment to build up a full picture of detail that gives you what you need to pull together the key elements of your data strategies. So, as you can see, we don't like waste either.

When you are looking at your data maturity assessment make sure that you have used it to completely understand the context of where you are, including the current drivers (including competitors, any external drivers and what is happening with your partners), business strategy and any regulation that will have an impact in the data area.

You must understand where on the risk adverse versus value add the organisation is and also where you need to be, as often they aren't always the same thing. Companies are very forward facing and are looking to maximise profit. This is not a bad thing – it keeps people in jobs – however, it does mean that when asked about what they want they will tend to focus on profit or make themselves look good. It can be useful to remind people that saving money also makes money too – don't be afraid to target the bottom line. It is useful here to return to the Henry Ford from Chapter 2: 'if I had asked people what they wanted they would have said faster horses'. It's a phrase that we don't use a lot but one that genuinely will help you, get your basics right. Like Ford, we are all just trying to get from a to b. Your data house won't last long if it hasn't been built on firm foundations so although in the short term you may look at the risk adverse stuff while keeping an eye on value add quick wins, don't let yourself be pulled too far in that direction from the beginning.

Workshops and interviews

The rest of the process involves pulling the company's thoughts together. A chart format is a simple way of combining all the myriad of detail you need to draw from. The second step in the process is to complete the chart in Figure 4.5 using a series of workshops and interviews.

Make sure you get a cross section from the whole organisation and that each area is represented. If you are not careful you can end up doing workshop after workshop and get stuck in an infinite loop. Have a limit on the number of workshops – as a rule of thumb you should be able to represent the whole company with less than ten.

In order to maximise your time and be efficient. Workshops are better than interviews but there will be people that just can't fit in with your

		Pain points	Initial changes	Ambition
	Strategy			
	Governance			
	Risk			
	Organisation			
	Framework			
	Policies			
	Leadership			
	Behaviour			
	Skills			
	Architecture			
	Metrics			
	Technology			

FIGURE 4.5 Chart for data strategy development

schedule so make sure you stay flexible. If it is fundamental that they are part or the process (and for some senior stakeholders it will be absolutely fundamental) then offer then the chance of having an interview with you directly. This isn't as time efficient but is worthwhile when you are in a corner. As well as covering each aspect of the company, make sure you have at least one area for the support functions, they underpin what everyone else is doing but often get overlooked. They do, however, use a great deal of data to help everyone else, as well as being big producers of data, so they will have a part to play in the strategy.

Use the workshop to explain why you want a change and what it is you are hoping to achieve. Always think about how you are engaging people in becoming data cheerleaders. Try to change the dynamic in the room. When you pull people together to talk about strategy, they tend to think in terms of what the next step is, they are limited by their current awareness of what is possible. So, mix things up by talking about the distant future, what is the true art of the possible? Perhaps show them a video of some really exciting things happening with data, anything that helps them to start thinking in a different way. Don't expect miracles but do try for a systemic shift and if anyone says 'but we always do it like that', spend a little bit of time explaining why that isn't an excuse for not trying a different way. You are focusing on the future not the past.

Take each line in Figure 4.5 first, don't try and just work through the columns as you will get stuck arguing about the little details when at this stage you should be looking at the big picture.

Start with the far right column first, 'Ambition'. Keep referring to whatever exercise you did to get people thinking outside of the box and looking forward. Where do they want to take this? Where do they want to be? This is a focus on the visionary piece, pulling in innovation. Push them to go further and further because whatever their first answer is will probably be too tame.

Next move onto the middle column, 'What will make a difference?' These are the current pain points in the company and can be fixed straight away. This is the much more immediate future through which you can demonstrate that you can make a difference quickly.

Finally, focus on the left-hand column, 'Current position'. Now you know that you already have the data maturity model and many of the people in the room will probably know it as well but it is still worthwhile doing this column. This is about the perceived current position more than the actual position. What the organisation believes to be the problem can be very telling because you have to deal with what they perceive as well as the reality.

The last part of the workshop is to look at which area is causing the organisation the most pain so you can pick up on which priorities you should focus on.

As you go along listen to the key words that people are using and capture those as well, you can use them in your strategy.

The number of workshops and interviews you have carried out will determine how many charts you have to deal with. Take all of these and merge them into one. Look for the commonality but look out for any real anomalies as well. You are looking for consensus as much as possible so you might not be able to include the anomalies in your data strategy but keep a record of them for when you are moving forward. Have a good look at where they are coming from and what is causing the anomalies.

The next step is playing back the consolidated chart to the same group as much as possible. Although these follow-up workshops shouldn't take as long as the original ones, they will still take some time as you want to work through the chart and give everyone time to have their say.

The chart should now be more like the one in Figure 4.6.

		Current position	What will make a difference?	Ambition
Purpose	**Strategy**	Data strategy exists, not aligned with current business direction No roadmap or delivery plan to deliver data strategy	'C' suite engagement with data strategy Funding for data-related initiatives	Data strategy integrated with business strategy Data recognised as enabler of strategic business goals and prioritised in business planning
	Governance	Good governance where legislation/regulation exists, i.e. GDPR, Open Banking, limited for non regulated areas	Consistent frameworks across organisation to provide platform for improvement	Common framework across organisation suitable to meet business, legal and regulatory needs
	Risk	Risks relating to data clearly defined and reviewed No agreed action plan in place to mitigate risk	Initial action plan to mitigate immediate high probability risks	Roadmap to deliver data strategy created and aligned with risk mitigation activities
Method	**Organisation**	Roles and responsibilities inconsistently defined, increased clarity where customer data is held	MVP RACI for core data entities impacting current BAU activity or key change initiatives	Data RACI agreed, understood and maintained.
	Framework	Elements of Data Target Operating Model (TOM) exist in silos across the organisation, inconsistent implementation and application. Where exists often linked to focus business issues or systems implementation not future focused	Definition of simple target operating model for adoption to provide basics for data governance, aligned with current business priorities	Common data target operating model defined incorporating the flexibility to meet business, legal & regulatory needs
	Policies	Many policies created across the organisation No process in place to communicate and ensure policy needs understood by relevant teams Policy review cycle not adhered to	Audit of current policies & development of action plan to remediate	Simple, easy to understand policy framework defined, implemented and embedded.
People	**Leadership**	No executive sponsor to champion good data practices High level of change across exec team	Create stakeholder matrix for current Exec team, identify potential executive sponsor(s)	Exec level support and recognition of data as a strategic enabler Exec data sponsor identified and actively driving change Stakeholder Matrix defined and maintained
	Behaviour	Positive energy to solving the data challenge, without frameworks difficult to focus efforts. Data champions exist in organisation, operating in silos	Build data community to enable collaboration and recognise the champions in the organisation. Engage to develop common understanding & approach to address the data challenge	Active data community working collaboratively to address current data challenges and mobilise for future data needs
	Skills	Pockets of capability within organisation No skills matrix or formal development in place, no talent pipeline building further future risk	Engage with HR/Learning & Development teams to identify competency frameworks for key data roles Identify existing skills across organisation & integrate into data community	Appropriate training and development aligned with data TOM & RACI
Tools	**Architecture**	Physical data models exist for all applications containing customer data, limited understanding of non customer related data No data ownership, no asset register No common standards for modelling data & information	Define minimal standards for data modelling & use on current change projects Introduce repository for storage and reuse of current data models Consolidate current view of information & data architecture Create community to maintain & develop data & information architecture	Common understanding of master data sources for core information entities Definition of data owners, data quality levels & data lineage
	Metrics	Data related metrics not included within corporate reporting. Monitored locally where operations necessitate	Identify how & where data impacts business performance, introduce into governance meetings to articulate the hidden data challenge	Data metric to be incorporated within Corporate Balanced Scorecard. Data inclusion in business performance monitoring
	Technology	High level of undocumented legacy solutions Limited integration between data sources High level of 'manual/spreadsheet' processing, limited automation	Work with IT/OT teams to identify high priority data sources, introduce governance across these core systems Local audit to determine key risk areas for manual processing of data	Removal of manual data processing, where feasible Common repository of core business data accessible as required to enable business strategy delivery

FIGURE 4.6 Example of completed chart for data strategy development

As you are pulling the chart together in a consolidated fashion you need to add in some practical steps that will move the company from one state to another. This is easier than it sounds when you are looking at the big picture. Just like the old phrase 'How do you tackle eating an elephant? One bite at a time', when you are facing the whole elephant it can seem really frightening but breaking any problem down into smaller pieces makes it more manageable. Taking it step by step gives you a list of much more practical steps to make a difference to each one individually.

In these workshops you are validating the consolidated worksheet to make sure that everyone agrees and can believe in the steps they need to take to fix it. This is where you are starting to build up buy-in from everyone.

Once you have this you have everything you need to pull both your immediate and target data strategies together. When you are creating your short-term strategy and have it bottomed out, remember to keep an eye on your longer-term strategy so you don't do something now that is wasteful – or at least keep necessary waste to a minimum and don't skimp on your quick wins (or minimum viable products).

These are the details that you refine first into your IDS – from the items you have learned from your data maturity assessment and content in the 'What will make a difference?' columns in your chart.

What is written in the 'Ambition' column provides the details for your target data strategy.

Vision

For each version of your strategy you need to just focus on what is important, don't try and put in all the detail from your chart. It is tempting to add more but you will end up over complicating it. The areas that you draw on to create the vision for the organisation are the business vision and the key words you have been pulling together while listening during the workshops. A vision is the guiding light – when we think we are going off track, it's a high-level summary of where we want to be that is simple and memorable. It has to inspire people now, so it should be written in the present tense, feel powerful and inspirational and describe an outcome. It has to build a picture in everyone's mind that evokes an emotion. Don't pick a bland statement that could be used anywhere.

Have about three supporting statements that underpin the vision, which add to the picture with a little bit more detail. After that comes action – demonstrating how you are going to change the end state, what is it you are really going to do.

Having the vision is critically important because it gives you the battle cry to lead your hearts and minds battle. However, without anything underpinning it, it can quickly become something of a joke. Part of your strategy is your high-level road map that demonstrates clearly how you are going to change from one state to another. At this level you shouldn't have a Gantt chart that goes on for thousands of lines, rather a guide to the key activities and critical line items that you need to highlight the direction of travel. At each state you should plan appropriately depending on how you are managing your transformation, whether it is waterfall or agile. But you don't need all those details right at the beginning of your journey (see Figure 4.7).

We often talk about looking at the two ends of the data value as being risk adverse at one end and value add at the other, there is no right place to sit forever and your attention should be more like a pendulum that swings between the two, making sure both spectrum areas are covered. Again, it's about balance, you have to at least cover both areas, but your focus could well be on one in particular.

Sample roadmap

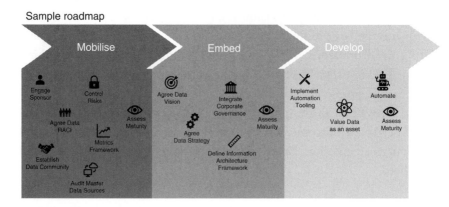

FIGURE 4.7 Mobilise–embed–develop diagram

Self-service

Ensure you cover your organisation's ambitions around self-service. Again, we are not advocating one particular path over another, we are encouraging you to ensure that you make mindful decisions and don't miss any areas of high potential. Self-service can be a powerful tool but it needs a high degree of data quality and control to make it work well. Are you ready for that or do you want to be? Making a decision about this area impacts so many other parts of what it will take to put in place your data strategy, from access to ease of use and the tools you need to put in place. It's a powerful tool. Think about a Wikipedia-like architecture for your company, for example, where your data assets are agreed and tracked, generating conversations about your assets. This is moderated so changes are tracked and agreed and you can create a self-sustaining, self-service model only it's not self-sustaining. It's one of those areas that looks easy but you don't see the army of things behind the scenes that need to be put in place to make it work well – such as up front controls, moderators and so on. Your strategy is not a place to be glib, think through each step of where you want to be and what the consequences of that will be.

Data science ambitions

The other area that you must think through carefully is your ambitions when it comes to data science. This is a massive area for companies – yes! Are you getting enough from your analytics yet? Don't just gloss over the power of your analytics, consider each step and carefully think through where you want to be and how long it's going to take you to get there. Your strategy has to be aspirational and forward thinking but if it is a work of science fiction as far as your company is concerned then no one will buy into it. Take it as far as you think you can and, once you have them on the journey and open to the possibilities of data, you can always revisit your strategy and power it up.

Keep the focus on this area, a data strategy is for life not just for Christmas. Refer back to it in your communication and engagement with the business, have regular touch points where you check it is really working for your company and that it's still pitched at the right level. Ambitious enough but not so ambitious everyone can jump on board with the idea. As your company evolves, are the answers to those questions the same? This is why you need to build in regular checks.

Data information ethics

The final area that we are strong advocates for covering in your data strategy is data and information ethics. What is acceptable to you? What do you think is acceptable to your customers and stakeholders? How would your actions be viewed? We aren't here to preach, it's about making mindful decisions again. Ask yourself if you would be happy to be treated in whatever way you are proposing in your strategy?

Within your thinking around ethics, also consider the unintended consequences conundrum. The whole point of the unintended consequences is that you don't see them coming and no matter how good your risk management processes are we can all be blindsided. So, if something does go terribly wrong, make sure you have a plan for how to fix it. Please don't wait for that something terrible to happen before you think of this plan. It's a waste of valuable time in a situation where you won't have time to spare.

Corporate governance

Felix van de Maele, CEO and Co-Founder, Collibra

Data is at the core of every organisation. To bring it to the forefront, an organisation has to treat its data as an enterprise asset and invest in building its foundation.

The right data governance and catalogue platform provides the foundation organisations need to gain competitive advantage by maximising the value of their data. Once it's in place, data citizens – the people in an enterprise who need to find data to do their jobs – can quickly find and use that data, while ensuring the quality of that data can be trusted.

Good governance within an organisation is very important. Unfortunately, the word 'governance' has negative connotations, people focus much more on what they can't do and forget about all the useful things that they can do. The value of actually having good corporate governance that also covers your data governance means that the rules of the game are completely clear. We advocate data governance being an intricate part of the overall corporate governance because the last thing you need to do is have people being confused by the rules of the game. In fact, making it worse, you don't want people believing they are playing different games or trying to play the same game with different rules. It would be very confusing if the referee on the pitch was following a different set of rules than the players. Even more worrying would be if three teams walked onto a pitch with one thinking they were playing football, one thinking they were playing basketball and one thinking they were playing hockey – that would just descend into chaos. Remember, the vast majority of people come to work because they want to do a good job – a large part of governance is making that easy for them so they are all playing the same game.

Knowing where you are in this phase of the transformation goes beyond just knowing where your data governance is. The understanding comes from how corporate governance is delivered, used and understood within your organisation. If governance is an evil word and people go out of their way to do the opposite then it is necessary to understand that so you can tailor your message around it. Don't create battles for yourself that you don't need.

Another aspect to consider is how people engage with governance. What we mean by this is, do people interact with governance on a very limited basis and only when they have to or do they see corporate governance as something that they can work with for the better? Working on any kind of governance role can be incredibly difficult because the hard part is understanding what questions to ask in order to get meaningful interactions with the departments. We have worked in lots of different places where departments will answer a question in a specific way to avoid interacting with the auditing department or any other kind of governance function again, rather than take it as an opportunity to fully understand the process they are going through. Being an auditor can sometimes feel like dealing with a very intelligent, creative three-year-old, they limit the answer so that they don't lie to you but don't feel compelled in any way shape or form to give you the full answer, which might expose the fact that, while they didn't have *a* chocolate bar, as you asked, they had *three*.

You don't want to boil the ocean with this area either. You want good governance in place across the organisation, so the focus here has to be on the data governance. The whole reason to understand the corporate governance is so that you can comprehend the impact that has on the data governance that you're putting in place. It isn't going to solve all corporate governance issues, just ongoing, to make sure you don't compound anything that already exists. The other aspect to consider is, if there is good governance in place already, why would you not work with the existing framework to incorporate good data governance as part of it? This will be familiar to the organisation and leapfrog you ahead of where you would be if you tried to do it in isolation. Never attempt to reinvent the wheel. Make it better by all means – make it different to be creative – but don't make work for yourself if you really don't need to.

So, what does the referee on the football pitch have to worry about? Firstly, it is important to be neutral, to make sure your data governance is coming from a place of neutrality. There's no point in having your governance viewed as a pet project for one particular area that can do no wrong. What this means is that you must make sure that as many people feel engaged with the governance as possible. And no, we don't mean that everybody has to mark everybody else's homework, what we mean is that they have an understanding of the overall process how it is executed – basically, everybody agrees that the referee is on the pitch.

Your mechanism for doing this is to have your governance documented, in a place that is easy for everybody to see and ideally agreed by your highest decision making around data and information. The whole idea of governance isn't a complicated one. In fact, your process around governance should be remarkably straightforward. At the highest level, is there a way of checking that your organisation does what is expected of it? That is all you're trying to do around your data governance. Try to keep the whole system as simple as possible, this is not an area to overcomplicate.

This could be part of your audit function. You could augment the standard audit function with the right questions and expertise with the questions around data and information within the standard audit that takes place; or you could agree with your audit function that a separate data audit will take place but work with them on the schedules. The one thing you have to make sure doesn't happen is that you provide data audits that are completely separate from anything the audit function is doing, otherwise one area of the business could be potentially overloaded by different audits within a short space of time.

The other key skill of a good referee is a full understanding of the rules. They must work to keep the game flowing. When it comes to data governance it's important that an expert has had a hand in helping with the assurance activities. And obviously they have to apply these fairly. In the corporate sense this is about making sure that there is a level of consistency that takes place in all of the governance assurance activities so that one particular evangelist doesn't make the test harder because they are incredibly passionate about processes, for instance.

Here, as in the vast majority of the rest of the book, we advocate a tailored and appropriate response to conducting audits within your organisation. It's highly doubtful you will have a full team of auditors capable of performing a complete data audit across the organisation in every department every single year. What is much more realistic is the idea of an overarching self-audit for the company, which covers all the main areas, checking compliance as well as understanding of the direction of travel for data within the organisation. As well as this, and depending on the answers from that audit, it would then be wise to look at a number of light touch audits across the organisation to validate some of the answers that have been given during self-audit. The final aspect of good data governance would be highly detailed audits on any area. We presume that doing around two of these every year would be more than enough for most organisations.

So, what you have is a three-step model that looks like a pyramid. What is covered at the base of the pyramid – the standard self-audit – goes across the whole breath of the organisation at a very light level, moving in a deep vertical strike straight through the organisation to understand an incredible amount of detail about a very small part of it: see the data assurance pyramid in Figure 4.8.

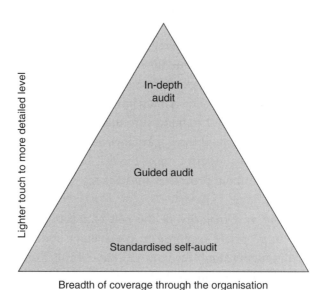

FIGURE 4.8 Data assurance pyramid

For each of these interactions or touch points with the organisation the same kind of questions should be asked to make sure that you have a level of consistency. This is where we should go back to the different elements of good data practice that are described within the maturity assessment and ensure that each area is at least touched upon within the audit questions. In that way you could end up with a red, amber and green traffic light system for each area within the department to give an indication of their current position, it will also let you look at progress in the future.

In fact, one way to look at it is this: with your audit you are reproducing your maturity assessments at different levels and at different times within your organisation. This gives you the ability to demonstrate progress across the organisation and in each department. Don't be afraid to celebrate the successes and congratulate the people and the departments that are really focussing on it – don't forget to say well done.

Once you have the right level of governance in place the next step is making sure something happens as a result of it. There is no point in spending a great deal of time within an organisation tying up people to fill in a form that gives you a red, amber or green light if you are not going to do something about it. If you're really smart the key to this whole area is being able to demonstrate and communicate the progress that is made as a result of the right governance activities taking place. Having the red, amber and green system helps you work with your department in order to look at what they need to improve in a particular area. We use those words very particularly, it is about working with them to improve, it is not about blaming them for not slavishly following something that they may not have understood. Playing the blame game doesn't help anybody; however, being a supportive challenging friend can progress the relationship to a point where departments are much more open and honest in their interactions with you so that really rich dialogue can happen.

Understand that these different departments have a vast quantity of pressures pulling at them from lots of different angles. Your role in helping them understand how improving the management of data can deliver value for them that makes it worth their while focusing on that particular area. In certain circumstances, however, this isn't the case. If you find yourself in that position having a level of understanding about the pressures that managers are under in that area can go a long way to helping your overall long-term relationship, but what needs to happen is that any decision to not look at an area is documented and revisited so that it doesn't get lost.

Using a colour-coded system like red, amber and green is also very useful because it gives a simple way of representing the different levels of maturity and compliance across the organisation. However, revisiting the numbers within the maturity assessment and utilising those can give you a greater level of granularity if that is what you need. Giving a graphical representation of the red, amber and green colours on the radar diagram means the problem areas really stand out.

Going back to the idea of something having to happen as a result of any audit or assessment, what should happen at the end of any good audit is that an action plan is put into place to focus on the areas that could do with a little bit of extra help. This is the kind of action plan that can be a quite light touch but have members from the audit or data team working with different areas ensuring they have what they need to be set up for success.

The ideal situation that we are trying to get to with all of this is that each department fully understand their responsibilities and what is expected of them, regularly checks that not only are they complying to all of this but are striving to improve and, if there is an area of concern highlighted, that they feel comfortable enough to be able to ask for help in resolving the issue. Once that starts to happen then you have absolutely cracked it.

Information risk

Kevin Fletcher, CDO, Director Data and Analysis, HM Revenue & Customs

There are so many clichés about the value of data to the modern organisation, it is becoming a cliché to refer to them. However, like any asset an organisation has, it has no value unless there is a clear strategy in place to understand it, manage it, curate it, exploit it and understand the risks to its use. In fact, without an underpinning strategy, mismanaging data and careless exploitation is likely to cause as much harm as good.

There is a ready tendency to focus on the shiny exploitation of data or the mystery of the algorithm to unlock value. This is important, but is only fully effective when underpinned by a comprehensive data strategy. That strategy will have three key elements to ensure effective release of value from an organisation's data.

Centred round clear governance that establishes data ownership, quality standards, regulatory compliance and oversight of value-driven exploitation, the foundation will see the implementation of clear strategy for managing data flows through the organisation. Authoritative sources for customer and operational data, a single data model and a clear strategy for acquisition of data sit at the core of an effective data strategy.

Layered onto that, an organisation must build its technology, people and cultural capability to manage, understand and exploit data effectively. A miner doesn't operate heavy cutting equipment without fully knowing how it aligns to its business process. The same applies to data. In many organisations, the data was not always treated as a prime asset, but advances in technology, tools and capability demand that organisations put these foundations in place for effective exploitation. Equipping staff throughout the organisation with the understanding, skills and capability needed to use this asset to its maximum potential is the second element of a sound data strategy.

The final element is driving advanced exploitation to unlock value. Releasing data to operational staff to service customers, giving data to managers to better manage processes, developing advanced algorithms to drive the business are all at the end of the process to ensure that true value is realised.

As much as we would like to believe it, a data strategy on its own doesn't add value. Having one in place mitigates the risks that data assets will not contribute to value. However, too many organisations do not understand this full value or manage the risks associated with exploiting their data assets. There are clear strategies that one should pursue to make sure that this is in place. We will name three.

First, the data strategy does not exist in a vacuum. It might seem trite to say so, but it must be embedded within the overall business strategy. When two strategies don't align, the business strategy will and should prevail. So, deep integration is paramount.

Second, the incentives and processes in the organisation must be such that the data assets can be fully exploited. If the data team is incentivised to manage, provide and exploit data but business teams aren't rewarded for exploiting it, the asset will be underutilised.

Finally, the culture of the organisation must embrace data as an asset alongside other assets that they more readily recognise. It is hard because data is intangible, even mythical to many in the business. To establish this culture, data teams must build trust, show value and embrace the business drive.

We talk about understanding where you are on the risk and value scale as being intrinsic to your data journey and how you move forward on it. However, you cannot ignore either side. The value side depends very much on what your business is trying to do and the mechanisms you set up to support them. But there is more structure to understanding your information risk and, while it might not seem like the sexiest thing you will spend your time doing, it is critically important and will help with understanding some of the priorities that you need to factor into your planning.

Any risk is a level of uncertainty in your organisation, so identifying your potential problems means you can plan for them occurring and close any potential gaps to reduce the risk. By mitigating any adverse effects on the company hopefully before they happen, this will help you avoid any hazards that can hit you and minimise their impact if they are unavoidable.

The reason we call this part of the transformation information risk rather than data risk is because it needs to be all-encompassing across the business. It's a little bit of a misnomer that we talk about information risk as well, it would be much clearer if we talked about information risks – plural. A bit like the journey with the policies, they need to be clear and focused on one specific risk – too wide ranging and you won't mitigate that specific risk. The medication is always much more effective when it is targeted. That said, there are lots of different areas that will need to be addressed. Each will need its own specific medicine but they will need to work in conjunction with each other. The last thing you want is for the medications to cause more problems because they fight with each other rather than try and heal the patient.

Why is understanding your information risk important? Too often risks are seen as a tick-box exercise – for data this is anything but that. It's critical to understand the information risks and the link between those risks, the impact that one has on the other, in order to understand the array of different challenges that could face the organisation going forward. You can then have a really clear picture of what you need to do and where your focus needs to be.

There are a number of elements to truly understanding your information risk:

- Clear short description of your information risk – this should be a very clear, concise description that explains the risk so that anybody reading it can understand it. Completely avoid the use of any jargon and acronyms, think about somebody outside your organisation, would they be able to understand this risk?

- Early warning indicators – what metrics are you going to put in place to give you an indication that you're moving into a danger zone? Don't worry if when you initially start this exercise you don't have the data to complete your early morning indicators. It's highly likely that you will be in a situation where you know what your early morning indicator should be but you may highlight a future early warning indicator and collect the data to the point where you can then usefully put it into practice. Remember that perfection is a disease. Put early warning indicators in place with indicative tolerance levels and monitor these to see if they are doing the job right, if they are that's fantastic, if they're not don't be afraid to modify them.

- Causes – make sure you look at internal factors arising from events taking place within the company and external factors, which arise from events happening outside your current organisation. Internal factors tend to be more controllable where is there is less control available from external factors. The causes could come from a wide range of factors, but remember to look at things around natural factors: competitive elements, any change – negative or positive – in demand force, better use of technology, human risk and changes in both internal and external control as well as the potential for any mismanagement.

- Risk assessment, safety, performance, finance and reputation (political) overall – hopefully your organisation has some kind of numbering system to assess your risks. Make sure you tie in with the corporate risk assessment when it comes to your data information risk and use the same system.

- Risk assessment rationale – this is the why section. Document the rationale for the assessment of the impact or the probability; it's important to document the justification for each impact and the probability of it happening in order to help communicate the significance of the risk. Likelihood – this is a pretty straightforward section. What is the likelihood of the worst outcome taking place against whatever framework your organisation uses? This will put it into the context of your other organisational risks and help you to understand the likelihood of it happening in relation to other important risks. This means that the organisation can appropriately tailor their assets and resources to the priority in the likelihood of the risk happening.

- RACI (responsible, accountable, consulted and informed) around risk. For all different areas within your risk, whether they are the overall area or the requested specific areas, you should understand who is actually responsible versus who is accountable. Whose opinion do you need to ask and who do you just need to keep up-to-date with what is happening?

- Existing controls, causes and consequences – don't assume that there isn't anything already in place. Ensure that you look for current controls that you can use to monitor this risk. Are there other processes already happening that the organisation still needs to complete that will impact this? Be honest.

- Improvement actions – what you need to do here covers two areas. First, try to stop the risk happening in the first place. Second, minimise the impact of the risk if it does happen and you have no way of stopping it. Both of these areas will form actions that will end up in your overarching roadmap alongside your other priorities. This is where understanding the balance between risk and value is important because you cannot ignore either priority. Make sure you put in place tangible actions with owners and dates. They should be simple, identifiable actions that are clear and that people can buy into. Don't assign owners if you haven't already agreed with them that they will deliver that action.

This is an area in which it is really useful to use the experts that you will hopefully already have in your company, in your audit and risk departments. You will be amazed at how often you go over the same sentences again and again, but this means the wording you end up with will be much more powerful and defendable when you start.

A good risk assessment is a living breathing document, it isn't something you carve into stone and then just occasionally view from a distance. Yes, you do need to spend time and effort making sure that it has the good preparation and coverage that you need. It also needs to be something that you use as a cornerstone of the transformational change of your organisation to make sure that, with all the changes that are happening and all the potential that could impact your company, you constantly keep yourself up-to-date. Different internal and external experiences, new knowledge, changing opinions from key stakeholders and advice from experts can all have an impact on how you treat your information risk. All of these things can be fluid, hopefully not too fluid, and as a result the information risks also need to be able to address this changing state. What you initially believed to be the correct way forward can be limited by your initial perception of the problem, as your capability and ability to deal with it grow, so should your ambition.

Chapter 5

People

L et's face it, people are a major component of any model when it comes to change – data is no exception. Skills, behaviour and leadership form the people-focused elements within the model for getting the data working for you. We have said it before, and we are sure we will say it again, it is so much more powerful when your whole workforce understands what a great asset your data is, and they have all the right tools for success, than a few specialists sitting in an ivory tower.

Skills

We wanted to pick up where we left off in our first book highlighted by Eden Smith around the skills we are looking for in data teams and the evolution of what is really happening in the market. Eden Smith have done some great work with Camden Council in the UK around how we can educate the next generation of children to prepare them for the future world in data.

Jez Clark, CEO EdenSmith

Skills in data – where are we? What have we learnt? What do we need to do to help, the facts?

2017–2018

In the *CDO Playbook* Volume 1 we talked about things to consider and how best to approach recruiting a CDO or Head of Data. At the time, this was one of the main questions on boards' minds. Since then many CDOs have been given the monstrous task of helping a business become data driven.

Today we talk about the skills and the development of those skills required to execute tasks set out by the CDO.

From a talent pooling point, we must understand that it's only the beginning of the journey, it takes years to build adequate pools of talent to meet market demand across multiple territories.

So, what is it we need to do? Who must come to the party?

The answer is, ALL OF US. It will take all of us and a huge amount of effort to build excellent networks that will deliver great things for the future in data.

Where are we today?

Unquestionably demand is still greatly outstripping supply. As we all start to discover the amazing things we can do with data, the desire and momentum to build teams of talented data people increases.

The main challenges are how we attract, build, position, deploy and retain teams within a market that's still shaping itself. There is just so much to learn, and many businesses have different opinions in how data teams should be structured within their organisations.

To put this in context, there are 2.5 million businesses employing people in the UK alone. Our view is that those who are not currently data driven will most likely be in the future. Hence, how will organisations find the people they need as they grow?

What have we learnt in the last 12 months?

Market alignment is key. If we all share a common view and a common language it will be far easier to build, attract and identify **roles** and **skills** within data.

Think about that for a moment. If you invited 20 people into a meeting to discuss the skills of a good data architect, opinions would differ based on experience and requirements of the idividuals' businesses. When you apply that thought to data strategists, data architects, data engineers, data analysts, data scientists and data governance experts across multiple industries, working for different people in different departments across different countries, you get a clearer understanding of why alignment can be so critical to the success and speed of building networks.

The message is, **keep it simple**. If you are a business interested in hiring a data architect then ensure that's what the job role is called when you write the specification. Likewise, if you are job seeking make it clear what your core job title and skills are. It is unsurprising that many businesses struggle to hire when trying to recruit hybrid functions and skills within a market that is still shaping itself.

What do we ALL need to do?

1. Work on alignment

    a. Speak the same language as much as possible

    b. Get support and advice from experts in data when recruiting

    c. Continue to collaborate as much as possible

    d. Be open and listen to those at the coal face of the market

2. Support academia

    a. Support the education system who train in data skills

    b. Work with organisations that train data talent

    c. Invest in training your own teams

    d. Partner with organisations that specialise in data

3. Understand the market

    a. The market is bigger than you – listen to the experts

    b. Understand what talent is available and how much is costs

    c. Re-invent your processes and hiring strategies to get results – get advice

Understand the skills required to fulfil roles in data

Research from EdenSmith showed very clearly the skills and attributes required to meet the demanding roles within the data industry.

The visualisation in Figure 5.1 was created because of the extensive work done with the STEM board, now STEAM for Camden Council, London to help understand what skills should be included within the national curriculum to build a future workforce of data talent.

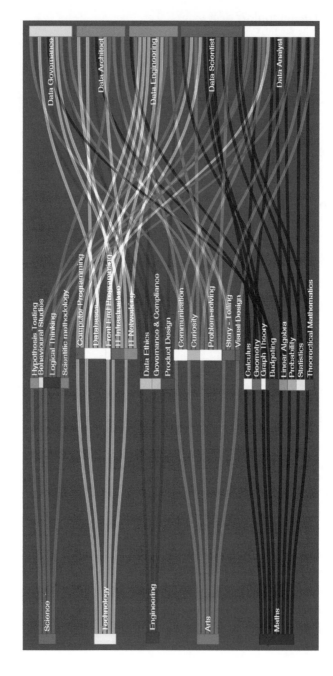

FIGURE 5.1 Eden Smith skills analysis

Used with permission of EdenSmith Group Ltd.

The insight came as a result of nearly three years of live job requirements and job-seeker data across the UK and Germany (see Figure 5.1).

It is safe to say that when we see statistics such as those in Figure 5.2, coupled with a view that things are moving faster, we have to all work together to build the workforce today of tomorrow.

This study by EdenSmith provides some great insights into the development of roles within the data sphere and just goes to show how fast

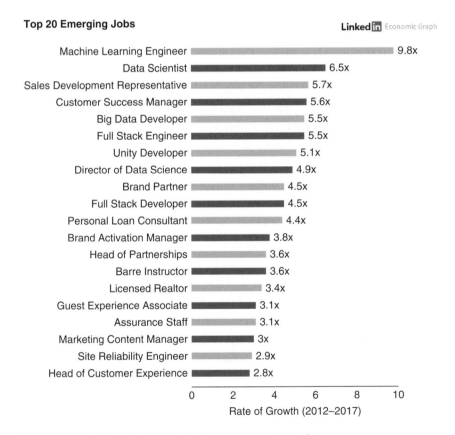

FIGURE 5.2 EdenSmith LinkedIn emerging jobs diagram

Used with permission of EdenSmith Group Ltd.

this world is evolving. It also highlights why it's important to assume that your data team will continue to evolve and why you must build flexibility into it, whether that is to use contractors or consultants for particular skills you only need for short periods of time or to augment the central capacity of your main team. However, you do need a stable centre that can build up your data capability and we go on to describe how this function itself needs to evolve.

Building your team will require different skills during its creation than when it is embedded. If you can get people who can morph and blend great, but you may need to think about a change 18 months in. This does happen quite naturally, as the initial problem solvers don't tend to want to stick around when it becomes business as usual. Obviously, this is a personal choice, but you need to be ready for when this happens as there will be a great deal of churn around the 18 to 24 month period. Relate back to the statistics on how much CDOs are moving around about this time – and they take teams with them. You will need to be aware of this of course. It can be scary because it feels like your intelligence is walking out the door, however, it is an opportunity to bring in the right kind of skills and experience to help you embed the change and take the organisation forward. This is also why your documentation must be top notch and fully agreed. History is important too, as you don't want to lose the reasons why something was initially put in place.

As you build up your teams and the skills within them think about the external roles that also impact data such as your suppliers, consumers (whether they are internal or external), modifiers and accessors. Consider the impact each position is going to have on these roles and make sure you think through the different roles and responsibilities to make sure there are no overlaps or gaps.

Whatever the organisational design there are some key skills that will be required in the data team, or in the business that the data team has access to. The more data-immature organisations will require an evolution of a data team and data skills that delivers the plan, builds and operates stages of the data-driven transformation. We will look at each of these in turn.

The plan phase

At this stage the skills required will be focused on data and information architects who will be required to plan and scope out the new data and information architecture of the business that will deliver and sustain the transformation. The architects will be creating the physical, logical and conceptual models. This team will also require business analysts to understand and map the data flows across the organisation in order to build the conceptual data model. The skills required will include data modelling and data lineage. There will also be a requirement for data technology skills, people who can understand the structure of the tools required for the proposed data activities; an understanding of extraction/transformation/load and the latest tools and platforms available. These people, inevitably, will be working closely with IT and the enterprise, systems and application architects. However, the organisation will need to learn and understand the difference between 'operational IT' and 'data, reporting and analytics IT'.

Some of the skills of the plan phase, and the volume of them required, may not be necessary in the later stages. It could therefore be prudent for an organisation to consider external resources to push through this phase. If this is the preferred option it is imperative that there is a full knowledge transfer from the external third parties to the 'home' team so that the intellectual property and understanding is available in-house with a full capability to support the ongoing transformation and the eventual end state.

The plan phase will also probably require a data governance specialist to talk to the company about data quality, data standards and data ownership. The data-immature organisation will likely be lacking in the areas of information asset registers and policies. Try to put your foundational elements in place to stop things getting worse. Up until this point your company has been creating an ever-increasing data problem. At some point you are going to have to tackle this problem so whatever you can to reduce the problem, or at the least stop it getting worse, will pay dividends in the future.

Of course, on top of the skills required for the plan phase there will be ongoing business as usual activities to support. The organisation will continue to require its 'data products', reports, market intelligence and business intelligence. Therefore, the skill sets for these activities must be retained and supported, and as the data-driven programme progresses these need to be 'brought into the tent'.

The build phase

The build phase will require the skill sets of the plan phase but perhaps on a smaller scale to match the level of build work that is actually required. Also, some of the build work may be being delivered by suppliers. Key skills required at this stage, on top of those of the plan phase, are without a doubt change management. Change managers will have to smooth the passage of not only changes in the data team and people associated with data but also the impact of the growing data capability on the business, which after all is undergoing a data-driven transformation. The change managers are often overlooked and need to be skilled in all aspects of people, processes, technology and data.

Quite distinct from the change managers is the requirement for project managers. Even though the data team, once established, will be working in an agile manner with scrum masters, the build phase will inevitably require project managers because there are so many more moving parts to be brought together that will build quite complex interdependencies between the data team, IT, procurement, HR, the line of business and perhaps even the legal department. Again, this is a skill set that may reduce in necessity during the later phases.

The build phase will require data engineers, that is, people with the skill sets to transform, blend and manage data. This is a skill set that will endure into later phases of the data-drive transformations. However, at this stage there may be quite a considerable 'hump' or high workload to get over, especially if the transformation is building and establishing a data-reporting layer, which has extracted data from operational IT and data warehouses.

The build phase will probably grow and diversify, or realign the skill set of the data analysts. The move may be away from management reporting, which should be becoming increasingly automated or set up in such a way that it may become 'on demand–self serve' by the department. Data analysts' skills set will be moving more towards business intelligence, advanced analytics, predictive and prescriptive analytics, data science and machine learning. Again, the blend of these skills will change as the transformation progresses. Indeed, the skill set requirement may evolve quickly as the data technology and platforms evolve. It is a fast-moving space and this is likely to accelerate as some rationalisation comes to the market place.

The data engineering team will be a constant presence, and again the balance between third party temporary resources and in-house capability is important. It is worth mentioning at this point that the whole thrust of the data Organisation Design and skill sets must focus on building capability. The needs and demands of the business will change, and at times rapidly. Innovation and disruption will become key words, which used increasingly and in order to meet these the data skills need to be seen as a capability.

During the build phase it will be important to have a DevOps capability. This is a function that has been around for several years in software development, but this type of function is very important to building out data capability especially when that capability involves new data environments, platforms and software. Increasingly we are seeing Dev Ops in this context being renamed, perhaps more correctly, as DataOps. The DataOps function allows data products to be brought to the line of business quickly and delivered out of proof of concept phase or Minimum Viable Product stage into operations. We will discuss the role and nature of DataOps in more detail in Chapter 9.

During the build phase it may be helpful to think of the data team as having three core elements: data exploitation data delivery and data governance. Data exploitation would include the data architects, data engineers, data analysts and data scientists. The data delivery team would include the functions of business analysis, project management, DevOps and DataOps and demand management. The data governance team would be focused on data quality, policies and standards.

The interplay and mutual support between the change management and DevOps is very important.

'Change management as it is traditionally applied is outdated. We know, for example, that 70 percent of change programs fail to achieve their goals, largely due to employee resistance and lack of management support. We also know that when people are truly invested in change, it is 30 percent more likely to stick.'
McKinsey (July 2015) *Changing Change Management*

In some organisations it is helpful to have the concept of a SWAT (special weapons and tactics) team, or a tiger team, that is, a multidisciplinary team with the full range of skills blended together to move fast to deliver a solution or outcome. The SWAT team may be a very useful delivery mechanism during the earlier stages of delivering a data strategy, while you are trying to mitigate burning issues and deliver some immediate business value. We discuss an immediate data strategy later in this section and in Chapter 9.

Throughout the build phase it is important to establish a framework for demand management. This requires the structures to interface with the business but also the methodology to assess the organisation's need in terms of delivery capacity and capability, cost and risk. This methodology should be endorsed, understood and accepted by the business. It will not be unusual for a growing data function to be inundated with requests and requirements from the business, on top of their own priorities, and this demand has to be managed in a transparent and consistent manner.

The need for training will be a feature of the build phase. Existing team members will require training to meet the new challenges and demands. This training has to be delivered with pace and at an appropriate level. Therefore, it does make sense to create some level of training management and delivery within the build team. A lack of training can adversely impact the quality of deliverables in the transformation either through lack of skill or engagement.

Even though you may have established a formal organisation design based on the three pillars of data exploitation, data delivery and data governance it may be worth considering the use of squads, tribes, chapters and guilds as an 'at work' methodology during the build phase. Much has been written about these but it is worth just highlighting the main principles here.

The highest level and most enduring is the chapter, that is, people or functions who do similar work. So, in our example, the chapters are data exploitation, data delivery and data governance. The heads of these pillars would be the chapter leads, who are responsible for developing people and setting salaries. Next is the Tribe, which is a collection of squads within a business area. The tribes are led by the tribe leader who is responsible for creating the right environment, through both technical and social and support services. The tribes therefore behave like an incubator for small start-ups and are the source of innovation. The start-ups are multidisciplinary, lean and produce the minimum viable products (MVPs). The squads are led by a product owner.

Finally, each individual is a member of one or more guilds depending on their skill set. Within the guilds they share best practice, knowledge, tools and ways of working.

This is a concept for organisation design, which is common in the world of start-ups and has driven Spotify in particular. If your chosen approach is a more hub and spoke design, or matrix design across the business, then the concepts of squads, guilds, tribes and chapters may be very useful.

It is vital to start building 'data skills' across the business during the build phase. A data-driven business transformation will drive and be dependent upon cultural change. To deliver this transformation the build team will need expert communications capability and the training capacity must be outward looking in order to build the necessary skills within the organisation. A currently high-profile in-house data skill is data literacy.

> *'**Data literacy** is the ability to read, understand, create and communicate data as information. Much like literacy as a general concept, data literacy focuses on the competencies involved in working with data. As data collection and sharing become routine and data analysis and big data become common ideas in the news, business, government and society, it becomes more and more important for students, citizens, (employees) and readers to have some data literacy.'*
>
> Wikipedia

'We hear a lot these days about "data literacy," but few stop to think about what the term really means. At a time when data is viewed as the lifeblood of organizations, it's critical that businesses help their employees to use data properly. After all, spending on big data and analytics products is expected to eclipse $200 billion by 2020, up from $150 billion last year, according to IDC [International Data Corporation]. To get the most from these investments, companies must educate their employees to use data properly.'

Doug Bordonaro (1 March 2018) *InfoWorld*

This ties into the ability to have a whole army of people treating your data as a corporate asset. Increasing the overall basic data literacy across the company means that not only is everyone speaking the same language but they understand the risks associated with data and what they should do about them (i.e., treat the data with respect, especially personally identifiable data). They are also constantly looking for ways to drive more value from the data because they understand its capability. Finally, they can interact with the 'data teams' more easily.

Again, do not forget that the data function will need to be continually supporting business as usual outputs throughout, but this in itself will be changing as transformation is delivered to the business through data. Don't assume that what works at the beginning of the process is what will continue to work as the company evolves into a great data-enabled company.

Run phase

The run phase will be very different from both the plan and build phases. This, by definition, is the end state of the transformation. This is a huge subject and, because it needs some considerable detail in its own right, is covered in its own section in Chapter 9, 'Running the business in the new data driven world, arriving at the destination'.

Behaviour

'Understanding importance of values and behaviours is key to success for any workplace, it can have direct impact on organisations' success, reputation and cohesion. When working with data in the workplace it is equally or even more important to have the right behaviours in place, in senior leadership roles such as a Chief Data Officer (CDO) emotional intelligence skills are essential for been able to manage influence, relationships effectively and been able to apply a story telling approach. Data professionals need to be able to speak the right language and bridge the gap between what Is perceived as 'geek speak' and turn it into conversations about 'art of the possible' or in other words speak the right language.

But aside from bridging the understanding gap behaviours need to work both ways (from colleagues working within data or otherwise), we need to be able to build an environment where we can navigate situations and make decisions – a workplace that understands importance of right behaviours and rates intelligence equally to emotional intelligence will be able to succeed in becoming data driven as it will have the right culture, values and behaviours. Whatever the sector or organisation industry we look at becoming data driven will soon be a given although a vital component of this would be to adopt the right values and behaviours.'

Sudip Trivedi, Head of Data, Analytics and Connectivity London Borough of Camden

It is very important to distinguish between culture, values, ethos and behaviours. Many organisations talk about these at a corporate level and expect these to be adopted every day throughout the business. So, an organisation may say that it has a data culture, or that it values its data. But cultures can be good and bad – just think of moulds and bacteria, some you need, some you would rather avoid. In a transformation we need to focus on behaviours, both corporate and individual. Having said that, what is most important is the individuals' behaviours because these aggregate to being corporate behaviours.

So, what sort of behaviours are we looking for in a data-mature organisation and how do we get there?

The first behaviour is data value, appreciating that data has a value. If you wish to acquire, buy or collect data this costs money. Storing data costs money, processing data costs money. Hence, the more data you collect, store, process and refine the more value it has, the more it has cost you. If that data then provides actionable insight it then has additional value. Once an organisation and the individuals within it start to demonstrate the behaviour of data value they realise that it is worth looking after their data – the valuable resource that they have collected and refined.

How do we encourage that behaviour? Well, first, we have to dig out the numbers that show that money has been spent in a data process. In other words, create the value chain of data in a business process or decision-making process, and make this transparent to the organisation. This is an important part of the communications and education involved in the transformation.

Second is valuing data. Creating a new sense of what you do is important. What happens when your culture values data as an asset? Even though we have just spun the words around, and it could be easy to get confused, it is a slightly different behaviour. This is when individuals seek out data as a valuable resource to help them in their processes and decision making. Valuing data is when individuals make decisions, or take action, based on data. This can be more subtle than it first appears. A data function may provide insight into a business process using the data and often, once this has been done and assimilated into the business, it is not the place of the data function to take action based on the data. It becomes the responsibility of the business to take action based on this insight. But how often do we see businesses fail to take this action? Perhaps because they are too busy or it would disrupt 'the way they have always done it'. In this case, the business unit is not valuing the data or the insight provided by the data.

It is essential to have accountability and responsibility around data and the actionable insights that support the right behaviours to value the data. A mature data organisation will have these accountabilities and behaviours adopted as second nature.

Governance is a very important behaviour. We draw this out as a behaviour quite separate from the frameworks and policies that we might talk about elsewhere. Governance needs to be embedded as a behaviour in a data-enabled organisation. The ultimate embedded data governance behaviour is fixed in a business process by block chain, perhaps because individuals cannot be trusted to behave in a governed manner. Well, not every business process in an organisation will turn over to block chain, many may do especially where external regulation or self-regulation is required. However, in most instances, individuals need to behave every day in each small and big interaction with data in a 'governed' manner. They must also appreciate the value of this. A very simple example might be individuals stopping themselves before they take a cut of data and export it into a spread sheet to carry out analytics and 'creating a new data set'. Governance is also a behaviour of ownership of the data, people in the line of business being responsible, and wanting to be responsible, for the quality standard and use of their data. This behaviour extends to the owners of the data maintaining and having full input into the information asset register for their data and regularly assess the maturity of their data.

A characteristic of a data-mature organisation is that it makes its decisions based on data, or on the insight provided by the data. To achieve this the organisation needs to serve up the data and analytics in a timely fashion. If the data isn't available then the organisation cannot adopt this behaviour. Basing decisions on data has to be habit forming, or perhaps it needs to be habit breaking. To explain this, data-immature companies will often base decisions on gut feeling, or what they always do, or experience, or we tried this before and it did/didn't work. Organisations need to break this habit and adopt new behaviours by 'trusting' the data. This trust is not only based on the more technical deliveries of data management and governance but is also built on changing the data culture. Quite often we have found that organisations may say that their data is rubbish or not good and shouldn't be trusted when, in fact, upon closer inspection their data proves to be reasonably good and could be trusted. The starting point for generating this behaviour is to change deep-seated misconceptions. Fear of the new can be a blocker to adopting this behaviour – like turkeys voting for Christmas. Individuals can feel that their role or job is threatened by the organisation's trust in the data and the fact that decisions are based on the data and insight it provides, which often will be automated, rather than on their considered, experienced, professional opinion and time-tested methodology. To shift this 'legacy' behaviour into a data-mature behaviour it is important to win the hearts and minds of these individuals – the turkeys – and demonstrate to them that often basing decisions upon data, data insights and data automation will release them to do more interesting, higher-level and professional work. Trusting the data and the data processes will enable them to focus on the important parts of the data or business operation, the anomalies, the instances of customers, operations or the business that do not fit the norm.

Data-mature companies that have undergone a data-driven transformation demonstrate collaborative behaviours. In simple terms this means that they don't operate in silos, they share data, methodologies and insights across the organisation. In a data-driven organisation collaborative behaviour enables them to move faster, to be agile and deliver business value quickly. It is interesting to note that very data-mature organisations collaborate with other organisations both within their own sector and across other sectors. They see the benefits of sharing data, being open and transparent with it. A data-mature organisation is capable of being open with their data quite simply because they understand the data and know which data sets may be shared with which partners or which data sets may be released as open data sets. A data-immature organisation does not demonstrate a collaborative or open approach to data, either internally or externally, because they do not trust their data. They are afraid of the potential consequences of being collaborative and open because they don't understand their data. So, in conclusion, perhaps the true measure of this collaborative behaviour is the level of collaborative and openness that the company demonstrates.

Innovation, or innovative behaviour, is a behaviour of a data-mature organisation. The company is prepared, ready and able to innovate within its business based on insights provided through data. It may even be that they are looking for innovation in the data. They may actually be asking 'questions' of the data in order to drive innovation. However, linked to other behaviours, the next step of innovation is to scale up and put the insight into action, into the line of business or business process.

It is clear that a number of these behaviours stack up together. An organisation that values its data, makes decisions based on data and is collaborative is very likely to be one that demonstrates innovative behaviour.

We will discuss in Chapter 8 the importance of agile behaviour in a data-mature organisation for delivering data-driven or data-enabled business transformation. Companies can only be agile if their processes, technology and people (and data) are set up and mature enough to support agile delivery. Linked to this agile behaviour is acceptance of failure. It is important to try innovation, to spend time looking into the data, but this won't necessarily provide benefits for the business, it may 'fail'. But if the organisation doesn't try they will never know. As many commentators and writers have said before, the important elements are to fail fast, to learn from failure and to value it.

There are two final behaviours that data-mature organisations demon-strate. The first is openness, this is not the same as collaboration though of course there is a link between them. In this context we mean that data-mature organisations are prepared to ask for help, both internally within the organisation and from external partners. Openness is a mature behaviour.

Second, is ethical data behaviour. A data-mature organisation will understand the statement 'just because you can doesn't mean that you should'. A data-mature organisation, for example, will not only observe and 'comply' with the requirements of GDPR, they will adopt the spirit of the regulations. They will respect the personal data of their current, past and potential customers and employees. The data-mature organisation will fully understand and have documented their data automation processes, they will fully comprehend and be able to explain any automated decision making or algorithms.

In conclusion, there is a raft or behaviours that organisation need to adopt, develop and sustain to be truly data mature. These behaviours are separate from the hard data skills that are required, that is, individuals who have the specific hard skills in data management or analytics. These are behaviours that must be adopted across the organisation to a greater or lesser extent. Perhaps this is what M&S see in their data literacy initiative? Establishing new behaviours will not happen overnight and will not be brought about by a top–down approach. Each individual needs to understand the value of these behaviours to the business.

Leadership

Jo Coutuer, CDO, BNP Paradis Fortis

Why CDO leadership matters

Thinking back to the agenda of several international CDO conferences in 2016, a frequent topic of discussion was 'the position of the CDO and his reporting lines'. In 2018, the topic is less frequently on the agenda, probably because it is a bit worn out as a conference topic, but not because the debate has been settled or the ideal solution has been found.

I would like to address three aspects of leadership surrounding the position of the CDO. The first is the leadership you 'get', the second is the leadership you 'project' and third is the leadership you 'deserve'.

The leadership you 'get'

The leadership you get is the leadership that is written in your job description. In my experience, CDO job descriptions roughly come in two flavours. The first one is the one that comes with expressions such as 'you influence', 'you drive the agenda', 'you enable the strategy', 'you convince', 'you inspire'. The second one is the one that comes with requirements such as 'you are accountable for', 'you manage the budget of', 'you decide about', 'you have operational responsibility for'.

Both are valid in their own right, but it is essential to understand the consequences of the 'begotten leadership' for yourself and for your ambitions.

First, you need to make sure that you get the tools to live up to the expectations that the organisation sets for you. A classic dangerous cocktail would be the following example: 'You inspire the top leadership to find new sources of income. Your success will be measured by how much new revenue the business functions generate.' If you are not in a position as a CDO where you decide which sources of income to prioritise over others, which investments to make and how to organise the go-to-market, then you can be as inspirational as you want, you cannot be held accountable for the ultimate revenue. If you cannot be accountable for the result, how will you measure the success of your inspirational efforts?

Second, you need to have the hierarchical leverage that matches the expectations the organisation has for you. In case of a CDO mandate that is mainly inspirational and influencing, the hierarchical position is less critical. However, beware not to be in the hierarchical shadow of another role that will not allow you to shine. Being innovative and inspirational as a CDO often requires questioning the practices of the past. If you are reporting to the traditional CIO, it will be hard to question the decades of data neglect that is typical to most organisations. For a CDO mandate with clear managerial accountabilities, the accurate hierarchical positioning is easier to assess. The importance of the accountabilities should match the level of the position. A good practice is to make sure that your accountabilities (often expressed in 'budgets') are in line with those of your peers. This may require the reshuffling of activities that were before in IT (such as business intelligence or data engineering) or spread across the business (such as data science) to give the CDO the levers he or she needs.

The leadership you 'project'

In order to 'project' accurately, you need to be comfortable with a good articulation of the position you 'got'.

If you have a CDO mandate that involves protecting your organisation against the risk of regulatory non-compliance, then do not hesitate to project your authority in the same way that your risk or compliance officer would do. Do not mix up your 'inspirational side' with your duty to 'protect and govern'. Install the same monitoring mechanisms for data as you have for finance, risk or business governance. Or better yet, do not install new mechanisms, but infuse the existing mechanisms with data governance aspects.

Dare to project yourself in different capacities. The CDO as governor, the CDO as innovator, the CDO as business generator, the CDO as technologist and so on. When you have to operate in different CDO personae, make sure to create the fitting contexts so that people understand which persona you are taking. Protecting against risk may momentarily seem at odds with driving for innovation. Be confident that, in the long run, only companies that manage to innovate in a compliant and sustainable way will prevail.

The leadership you 'deserve'

A CDO, regardless of his or her hierarchical position, best operates as a servant leader. Even with clear operational accountabilities, the CDO is ultimately an enabler.

If you organise your operational activities in the field of reporting, business intelligence, analytics, artificial intelligence and so on in a way that is (internal) client oriented and (external) customer relevant, then it will be easier to achieve your objectives in the governance roles that you have.

On the other hand, if you treat the topic of data governance as a topic that needs to be 'sold', you are at risk of not fulfilling your protective role. After all, if data governance was saleable, it would have been done better in the last 50 years. Likewise, if you believe you can impose technical standards, service levels, product standards and the like on your internal clients in a 'govern-ator' style, then the organisation will find ways to bypass you, with external suppliers being the easiest avenue.

Conclusion

With a relatively new position as that of the CDO, the key success factor is to carefully develop the equilibrium between the position you get and how you project it, while at the same time making sure you show that you deserve it. For other functions, such as that of the CFO, CIO, CHRO, these dimensions and their balance have been more extensively described, practised and accepted. Every CDO out there is contributing every day to the shaping of his or her profession. That is what makes it exciting to be a CDO in these times!

There are two parts to this section on leadership. Firstly, there is the leadership that one person can bring to the organisation when they are the central data cheerleader. As Jo Coutuer describes so eloquently, this is a really exciting time to be a data leader but that doesn't necessarily equate to it being an easy role. It takes a certain kind of individual and that is coupled with the transformational aspect of the role in the first place. When we are being kind we call it pioneering, when we are being not so kind it has a more of a wild west feel to it! While we want everyone in the company to treat data as an asset and use it accordingly, in conjunction with the strategy, your CDO, director of data or whatever you decide to call them needs to be there to provide that central focus – to raise the flag and sound the rallying call for everyone to march behind.

Due to the different emphasis between the transformational and the business as usual the person who leads through the change may not be the best person to lead through the business as usual. They could be, and we would' want to rule this out, but don't keep trying to push a round peg into a hole that has become square. We explore this more in the organisation section in Chapter 6.

What we mean, when it comes to the other aspects of leadership, is that it's so much more than whether or not you have one person sitting on the board who understands what it is you're trying to do. Of course, it's great if you do have someone who gets it, and sometimes one person can make all the difference. We have worked with organisations where that one person can turn out to be a key and critical difference in how the changes are embedded within the organisation.

An example of this is the work we did with a really large organisation in which the chief executive was very committed to the idea of change. In order to try one particular aspect of the change the executive in altered their working style. This had a knock-on impact that went through the whole organisation. Quite simply, rather than regularly sending out documents and presentations to lots of people across the company when requested, they began to store them in a central area. Hence, every single time they were asked for a presentation they could direct people to the correct procedure for accessing it and ensure that everyone was receiving the correct information from the same place. As you can imagine, that influenced everybody else's behaviour around them and, as they became used to acting in that way, it set the standard for other leaders throughout the organisation to follow the same procedure. So, one person can make a big difference.

Trying to implement an organisational wide data-driven business transformation without committed leadership and sponsorship in place puts you in an incredibly difficult position. You really would be pushing a massively heavy boulder up a never-ending hill. It's natural for senior executives to have an area a primary area of focus, we would expect, for instance, your CFO to be pretty dedicated to what's happening with the numbers. At that level they also need to be committed to the strategic direction of the organisation. Your role in all of this is helping them understand how important the data-driven business transformation is to the overall strategic direction of the company so that everyone is committed to that goal as well as their main individual focus.

Even worse is if you end up with a data villain sitting on the board who could be incredibly damaging to the programme overall. Let's look at an example from the opposite end of the spectrum to the excellent data sponsorship created by the chief executive of the previous company. A very senior stakeholder on a board was on paper a great sponsor of data and how it would be used. In fact, on first impressions we believed they would be one of the biggest sponsors. Disappointingly, rather than being a key sponsor, they were actually incredibly good at undermining the data programme. When we finally understood what was going on it was a matter of politics rather than any real lack of commitment to or understanding of the data transformation. That is very poor behaviour for any senior stakeholder or leader within an organisation but, unfortunately, it is something you need to watch out for. Dealing with any potential rumours and ripples caused by this kind of situation quickly and efficiently means that they aren't allowed to go on unchecked.

Hopefully by now you have a really clear understanding of where the board and other senior stakeholders sit in relation to their commitment towards the end date of business transformation, it's an area that you can never take for granted. The engagement that takes place at this level is never ending.

There isn't a simple step-by-step guide to what you need in this area. What we can give you is an extremely high-level overarching process to follow. When you're dealing with very senior stakeholders who all are responsible for complex areas and levels of engagement, each one needs a tailored action plan. If you haven't already got a list of all the senior stakeholders then now is the right time to make one. Ensure that the list is not limited to just the board members – there will be very influential people within your organisation, who have gained that influence either through their expertise or longevity of service, to whom people within your company listen, they should all be included as senior stakeholders.

It can sound a little over the top but for each of the stakeholders you need an understanding of whether they are supporters or detractors, what you believe is causing them to be that state and any actions you can take to move them into a more positive state of mind. Literally, we are saying that you need an action plan for every one of your senior stakeholders. In lots of cases this can be the same action plan for a number of different people, but it should still be very relevant to them. This document needs to be updated at least monthly so that you can track the state of mind of each of them. One word of caution: do not forget your data heroes and those who fully supported you at the beginning. It's incredibly easy, and as human beings we think it is hardwired into us, to focus on the negative and forget about the positive. The last thing you want is to turn a great supporter into a detractor because you haven't paid them enough attention.

Of course, you want to try and standardise the meeting process as much as possible while still making them feel tailored. Getting regular slots in already existing meetings where you can continue to demonstrate the progress and highlight any areas where you need the help (everybody likes to feel wanted – just don't make it for anything unnecessary) means that you have regular checkpoints and touch points with the senior stakeholders. This ensures that there is never too much time passing without some kind of engagement. The world is a busy place, a gentle reminder that your programme is there, continuing to change things and make a difference within the world of your organisation does a lot of good, it won't harm your programme either.

When you are discussing things with senior stakeholders make sure that the focus continues to be on the big picture. Always have the mantra in mind of what's in it for them when you're discussing it – but make sure it's the big picture version of that. Using individual examples where you can highlight the what's in it for them for individual departments, where real changes are starting to happen, is a great way of dropping something into the conversation, just remember to keep pulling things back to the big picture.

A great deal of the success of the transformation depends on the commitment of and consistency from your senior stakeholders in continuing to back the programme through the entire process. It's your role to make sure that they understand what is happening so that nervousness isn't allowed to creep in. With an air of confidence in the change as being a good thing for the organisation, it's controlled it's understood and there is a clear direction moving forward, which will make a significant and real difference to the organisation, delivering value in a way that the status quo can't.

Chapter 6

Method

With the 'method we incorporate the organisation, framework and policy elements of the model. This part of the transformation process is concerned with how something is put in place to help the company be successful. Basically, this covers the rule, the referee and the team when it comes to playing the data game.

Organisation

Graeme McDermott, CDP, Addison Lee

When Manchester City's new owners spent hundreds of millions of pounds on assembling the best players on the biggest wage bill in the world, they still didn't win anything for a few years.

Whilst I'd love data professionals to be on footballer salaries and retire at 35, I am a realist who wanted a career that lasted a lifetime. However, in most teams just grabbing all of the best resources isn't good enough and won't necessary lead to success. How you organise your resources and get them to work together as a team is often more rewarding for the team and cost effective for the company. My own secret for success is to group the data engineering team with the business intelligence and analytics function, along with the insight and science team.

Whilst they often use different tools, they should be using the same data and definitions. Keeping this group together allows you to balance data requests across the teams when resource conflicts exist, share data language and tips in the same area and create some healthy intra team data competition.

Now you've assembled your team and they are working together, where do you sit in the organisation? Does that really matter? Believe it or not, from the interviews I complete it seems to. The number of times I hear from data professionals up and down the ranks how a change in perceived organisational importance, either through directorate change or new leadership, can have a massive disturbance on the team. I have seen whole teams implode and the company's data reputation go backwards many years by failing to treat the data team with respect, after all they will have the data and models to predict what will happen next!

The degree of data maturity will often dictate what the data team looks like, or if there isn't a data team the level of maturity will determine the existing roles and responsibilities for data within the organisation. In a very immature company no one will be taking responsibility for data (see the information architecture section). There may be individuals dispersed across the organisation who produce reports, management information or business intelligence. There may even be small pockets of individuals carrying out some level of data science. But because these 'data' activities are not joined up they are immature. Equally, in a data-immature state the business will probably have abdicated responsibility for 'data', its quality, standards and governance to IT, who will be managing the data as an IT issue and discipline, probably focusing on the 'tin and boxes' or databases and systems rather than the data. And in the worst case scenarios, the IT function will not have picked up the responsibility at all because they aren't even aware they should be, believing that the business are responsible for the data. In some organisations when you ask who is responsible for the data everyone appears to just point to their favourite part of the wall.

Hence, that is an immature state, and of course there are graduations of immaturity to maturity. One of the most important things to take on board is that it doesn't really matter what the organisation design looks like or where responsibilities and accountabilities rest so long as there is full responsibly and accountability for data that is understood and respected across the company. There is a little bit more to this, which is rather important: it does not matter what this organisational design looks like so long as it is properly functional for the business to derive and exploit full value from the data and for the data functions to be responsive to the business needs and questions as they evolve. There is no point using a centralised model, for instance, if the whole business works in a matrix fashion, you will just build up a level of resistance that will cause issues as people try to adjust to a new (for them) poorly cut suit.

Moving from the start state of maturity to the desired end state will inevitably require some decisions about the roles and responsibilities and organisational design. The first options will often rest around whether there is a centralised 'data team' or data function or whether this data function is federated out across the business. Each has its own pros and cons. For an organisation that is starting on a data-driven transformation from a very immature state it may be favourable to build a centralised data organisation. In an immature environment one of the key elements to get hold of is data management and governance. The quickest and most effective way to achieve this is to bring together a centralised team that will quickly establish common standards and qualities for data (or at least common approaches), common data management techniques and standardised reporting and analytics. Bringing together a centralised team may also deliver efficiencies that may help build the business case for a reorganisation and swiftly drive the transformation. However, a reorganisation of people and capability can produce its own problems in change management and must be carefully managed and delivered. If this approach is taken then much of the CDOs early work may be on team transformation delivering a new organisation design and target operating model.

Alternatively, if the business is resistant to the level of change and/or risk that bringing together a centralised team may cause it may be more prudent to build a federated organisation design. A design more along the lines of a multi-hub and spoke model – a small central office of the CDO, focused on standards governance, policies and centre of excellence for data, with other data functions such as market intelligence and business intelligence and even data science federated out into the business. This is possible depending on the level of data maturity that is the starting point. One advantage of this approach is that it focuses the early work on the data rather than on organisation transformation. The aspect of this model that must be in place is the constant enablement of two-way communication. There should be data champions, that is, people who sit within the business who have strong ties to the data team. The data team works hard to listen to and work with its data champions, setting them up to succeed and using the information they provide to improve the data offering across the company. The data champions in turn, while they are full time in their day jobs (such as finance or operational roles), are committed to working with the data teams, helping them hone policies and working to achieve a high degree of assurance and constructive challenge.

Whichever model is adopted (and it is not as simple as just one or the other being the right one for your company – it will probably be a hybrid of the two) as the route forward, small responsibilities and skills will lie outside of the data team either with IT or in the business as a whole. A good example may well be information security and information management, which may sit in the IT function. Again, it really doesn't matter where these accountabilities are or the escalation route of responsibility so long as it clearly defined, understood and accepted across the organisation.

A central point of focus is a crucial element of setting up your data organisation. In our previous book we talk about the role of the CDO but we aren't precious about that title. You can call them data leader, head of data or general data overlord of the universe, what *is* important is that the data vision has a cheerleader who will focus all their efforts on making it happen.

Who is going to maintain a focus on efforts in the information value chain? How does the information you use link into the value chain for the organisation? What end result are you expecting and what do you need to get there? This isn't just about using the five whys – don't stop at five, start asking questions and don't stop. The focus has to be on delivery and benefit. If collecting the data doesn't deliver benefit, stop doing it. Keep your limited resources focused on doing what gives you benefits. The CDO (or grand overlord of data) gives you this clarity and direction for your attention. There is no point in putting in place a great team structure that works really well for your company without giving it a point of focus in a strong leader who is at the right level within the organisation to make a difference. Hiding your data leader and teams so far down your company that they are in the organisational equivalent of the basement will not demonstrate to anyone that you are taking your data seriously. How can they continue to maintain a focus on improving the value the company derives from its data if they don't sit at the strategic level?

Bear in mind that the data leader cannot be everywhere all the time; they will want to be and will need to be to get a handle on the business, a handle on the burning platforms, to stop more mistakes from happening, to start changing behaviours and culture and to start spreading the word. They need to be visible and at the same time they need to be managing up and across – they need to keep the data message going right up to the CEO and the board, (the board may expect this anyway) as well as fellow directors or other senior colleagues. So, early on a trusted data-savvy lieutenant is essential and quickly backing this up with a team that works together will help keep their sanity. Part of the organisation element of the transformation process is therefore about putting these elements of support in place.

One area that we are very clear on is that whatever model you use and however organised you are, to the business there is one data team. There is no point putting forward any structure that requires the rest of the company to become confused about what part of the data team they need to talk to. Structure yourselves in such a way to keep it simple (or at least to appear simple, you can always do a swan dance – serene on the surface and paddling like billy-o underneath). It is up to you and any other senior staff to cope with the intricacies of the model – not the stakeholders who will be working with you.

We want to make people curious about the data they have. As with a child, you never know what questions curiosity will lead to but it might pull up some really interesting answers to things you never knew you needed to know. Questions can drive the data journey forward so how do we enable people to ask questions and feel comfortable doing so? How can we tie all of this back into measurable outcomes (both good and bad, because you learn as much from your mistakes as you do from your successes, as long as you don't continue to repeat them). Let's stimulate the debate and take action in an agile, flexible way using the power of your whole organisation.

Whichever model you utilise you want to create communities across your business that can be connected and collaborate. However, they also have to be able to operate on their own when they need to. This helps with keeping the speed of decision making as close to where it needs to be as possible. Getting the right information to the right place at the right time is the data team's responsibility, so the rest of the business can use it.

It is becoming more common to talk about citizen roles in organisations (people like your data champions who have day jobs but are performing data elements within those roles, such as a finance person who is using analytic tools to understand and utilise the data rather than going to an analytics team to do it for them) and how these will grow as the maturity of the organisation grows. Imagine if everyone in your company understood the value of your data, worked to ensure the quality of it and treated it as an asset. You would have an army at your disposal rather than a small team locked in an ivory tower. This is an incredibly powerful model and one of the reasons to work extremely hard to win your hearts and minds battle.

The framework for how all the different data areas will be defined shortly. However, that doesn't cover the clear accountabilities across roles. As much as you want data communities all linked together (and you really do want this), you also want the roles and responsibilities to be clearly defined so that when they are interacting with each other they don't spend more time stepping on each other's toes than they do making a tangible difference to the business. Goodwill will only take you so far, make sure that all the roles identified in your model (see the skills section for more details) are clearly documented and agreed upon so each person feels empowered and not limited.

Framework

Mohammed al Madhani, Data Management Manager at Abu Dhabi Smart Solutions and Services Authority

Abu Dhabi is geared towards transforming government processes as part of its digital transformation agenda, including unlocking the value from data and its massive potential to spur significant and sustainable economic growth in line with Abu Dhabi's vision and strategic directions. The emirate works to enhance its data management programme to consolidate information under advanced open data platforms following the UAE's Artificial Intelligence Strategy and the latest developments in data analytics. Our goal is to enable efficient and effective delivery of services to help organisations and individuals make faster, easier and better decisions.

Data management is one of the core pillars that Abu Dhabi is building on to achieve its vision of a sustainable knowledge-based economy and society. The Abu Dhabi Smart Solutions and Services Authority works with various government entities, research groups, start-ups and the private sector to accelerate our programmes in data management in a secure manner to create the right data value products. We have advanced our efforts towards streamlining accessibility through data journeys ecosystem, while keeping data governance privacy and security at the core of our efforts and maintaining trust and reliability from our data sponsors and sources.

The key part of the framework is understanding how all the different pieces of your data organisation make decisions, it's the key understanding that helps the whole machine of your data lifecycle work within your organisation. As Mohammed Al Madhani describes very well, it's the many different pieces that all have to work together. We tend to think of organisations being giant machines with all the components – from the cogs to the flow –working together to make something quite incredible. The framework is how the machine works for your data in the organisation – we're not talking about data flows here or things that are covered within the information architecture, rather it's about the data lifecycle and operation in your company.

This is also one of the areas that a lot of other elements within the data framework rely on, the basics in this area need to be in place in order to help the other elements work. The major function of this area is to take care of the rest of the elements, it is the one relied on to make decisions about the data. Whether that's around the information architecture or policy being signed off or if it's a lead on how the engagement is focused across the organisation, this is the area that ensures the company has the ability to do those things.

From that it should be pretty obvious that the initial starting point for this area is to put in place a data council or information steering group. This will be the highest decision-making body across the organisation –capable of making the big decisions. It has to be representative of the whole company because it will be making decisions about the data for the whole organisation. It must be staffed with decision-makers who have the authority to make those decisions during meetings and do not have to refer to anybody else.

No meeting should exist without clear terms of reference. There is absolutely no point in having a meeting for the sake of having a meeting – these are just practical alternatives to work where people spend too much time talking and not enough time making decisions. The key to stopping this from happening is to have a clear terms of reference that people are not only signed up to but held accountable to. Your highest body should sign the terms of reference off at the very first meeting; everybody who is part of the meeting is then beholden to play their part.

A trick to understanding who should be in this meeting is to look at your information architecture, the highest level of which should be somewhere between the five to eight biggest data domains that cover all the data across your organisation. Understanding who is accountable for each of the data domains gives you the insight into who these key people who understand and are responsible for data across your whole organisation are – these are the people that should be in your data council (see Figure 6.1). The problem with this is that you will probably need to be able to make decisions around your data before you have those key accountable people in place. Your choice here is to either use the programme board for the data-driven business transformation initially, knowing that it will transform and morph into the business as usual data council, or you estimate who those key players are in the knowledge that the make-up of the meeting will potentially change going forward. Either way will work, it simply depends which will work for your company.

		Accountable for...
Headlines	Provide strategic direction, decision making and oversight for XXX's information governance	Information governance strategy Reviewing and highlighting significant deviations from the information governance strategy
Members	Chief data officer (chair) Representatives for: Operations Assets HR Finance	Review progress against the information governance implementation plan Assuring that information risks are properly identified, assessed, controlled, mitigated and contribute to reducing the level 0 enterprise risk
Meets	Every month in the initial phase: to be reviewed after six months Ad-hoc meetings may be called as necessary	Approving information governance policy and monitoring assessment of conformance across business functions
Quorum	At least four members: representatives are asked to send delegates if they cannot attend	Reviewing the support offered to business functions in implementing information governance best practice and conformance with the strategy and policies
Ways of working	Meeting agenda to be circulated at least one week prior to meeting date, papers to be submitted at least two working days prior to the meeting Definition and use of KPIs and KRIs to monitor and report on the performance of information governance related activities	Reporting to the legal and corporate services leadership team on the performance of information governance activities Prioritisation of information governance related projects

FIGURE 6.1 Example of a terms of reference for an information governance council

Typically, that's probably not the only meeting that you need in place because you can't have a maximum of eight people making every single decision about your data. A lot of the decisions will need to be made by different subsets of people who are either experts in particular areas or more focused and frontline in the area that they're working on. Breaking the meetings down into clear decision areas is a good way of building up your framework. Like a lot of the things that we advocate, start small with just one or two meetings and see if that works. If you find that you have one particular area that doesn't interest everybody within the meeting it may be a good idea to have a sub meeting or a breakaway meeting in order to get the participants focused. You really want to make sure that meeting attendees are focused on what you're talking about, but also that it is an area that interests them, that they have something to contribute to and gain from. Otherwise, if people's attention is wavering and they don't see the benefit of attending the meetings, you will get meeting drift, where they cease to turn up or they start sending deputies. Even worse you could end up with deputies of deputies so the ability to make decisions is lost from the meeting entirely.

At a minimum there has to be one overarching meeting as we have already talked about. You can imagine it being the one source of truth for the organisation, and it should also feed into part of the corporate structure. Depending on the main focus for data improvement across the organisation this could be part of your corporate risk meeting, it could go into your executive group. Wherever you put it, make sure that it continues to the focus on your data and information coming from the right level and that it gives the right visibility across the organisation.

All the areas to make sure that you do get covered all your information architecture and your overarching governance. Both of those areas, especially at the beginning, need a lot of feedback from the whole organisation and extensive two-way communication to make sure that the organisation feels and is being listened to – watch your delivery. Although they may not sit under your direct responsibility, part of this framework also has to encompass anything around information security and data protection. These areas are within the data ecosystem and there is no point in having three or more different frameworks that look after data – that is just a sure-fire way of confusing your organisation. We are big advocates of the idea of the matrix team within the data world in your organisation, it's not your company's problem if you aren't working in the same department as the information security or data protection team, you should be striving to make people across the organisation comfortable with data. Your aim is to make it so simple and straightforward for them to do the right thing that they wouldn't dream of doing anything else. So why would you create three different teams that they have to think about in terms of data information? What you should be aiming for is a one-stop shop and your framework reflects how that would come across to your organisation.

Figure 6.2 is a draft framework covering some really basic meetings and how these could work together. For each one of those meetings you must have a terms of reference. There is no point in having a meeting without a purpose – as we have discussed, this will just end up being a talking shop with no decisions or actions taking place. It's also important to understand the hierarchy of each of these meetings and how you keep each one updated with the necessary information to be able to operate efficiently and transparently.

The final point that we want to make in this section is understanding how you solve problems for the organisation. While the common core of your framework is static with regular meetings taking place, there are also meetings that should be stored up on a project type basis in order to solve specific problems. We're talking about the kind of activities for

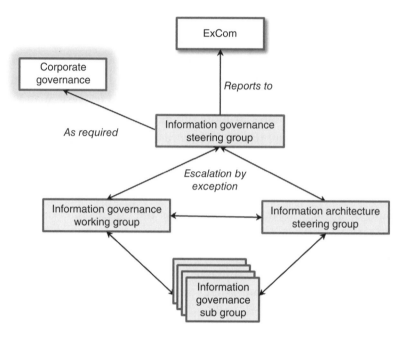

FIGURE 6.2 Example of a high-level framework

which it will help to use data champions and other interested parties who can come together to solve a problem and then be disbanded rather than trying to overlord small parts of the organisation. As the old saying goes, if you want something done you should give it to a busy person. We still find this to be true in organisations. We find that certain people, because they are so good at what they do, are asked again and again. However, this is a really easy way of overwhelming people and, if you're not careful, your reliable person can quickly become overloaded and everything comes to halt. Having meetings where you can use a collection of people in an agile fashion to come together to solve particular issues for a limited duration of time will spread the load through your data champion network. This also allows people with particular skills to bring those skills to bear when it's helpful and necessary but not have to get involved in areas that they have either no interest in or no ability to influence.

Policies

Stephen Gatchell, Head of Data Governance at Bose Corporation

Policies enable efficiencies by providing guidelines to data users around data availability, data usage and data sharing. Policies also provide transparency to data creators on how the data is going to be used. Managing expectations for both creators and users ensures surprises are kept to a minimum (meaning stakeholders know what data they can use, how they can use it and who they can share it with). Keeping in mind that policies need to be affected.

As we have previously mentioned, the policies section is concerned with more than just policies – it incorporates every control document used in your organisation. Different organisations play to a different tune but they will all have some documents that tell people what to do and how to do it (whatever *it* is!). Asset heavy industries tend to focus on standards from their technical background whereas companies who have grown quickly usually have a plethora of policies. It doesn't matter what they are or what they are called, what matters is whether or not they are working. If you are focusing on this section then we are going to assume that from your maturity assessment you have identified a bit of a problem. In highly technical language 'something somewhere has gone terribly wrong'.

This is an easy stage of the transformation process in which to make some major inroads into changes in your company and demonstrate what you are doing. The hard thing is that the outcomes that you are trying to achieve will take time and are much more difficult to demonstrate, but they really make a significant impact. This is the difference between having an army marching to the same tune with a common objective or a pile of conscripted farmers who all think they have something better to be doing.

First things first, get a copy of every one of the policies, every single one. They are really good at hiding in small sections of the company so make sure you do a good sweep. Especially if you have had a set of data heroes working independently in silos in your company, this tends to be an area they have started on but probably won't have collaborated with anyone else on (normally because they didn't realise other people were trying to solve the same problems). So, pulling together everything is a really good starting point.

There are two main issues you are trying to solve here: one is to get a set of control documents that work (more on that in a minute) and the other is a means of everyone in the company being able to access them. We will tackle the first issue first and then talk about the second.

Getting all your policies together is a good starting point but you still have a lot more to do. Read them and see what your initial impressions are. Policies grow where people see a need or have an issue, so if you are overwhelmed by policies around security then that is something that needs a strong focus. If what hits you in the face is a lack of anything around data classification then does the company understand the importance of something like that? Which areas of the business have the most policies? They are probably the areas that will either (a) stick to their guns and say that they have the most wonderful suite of policies ever and everyone else just needs to follow their example (which might be the case but should be put in more tactful way) or (b) understand the value of getting policies that work for everyone and will be your biggest supporters in this area. Only by talking to them can you figure that part out.

This is another one of those areas in which you can't let silos get in the way. If you don't have full control of all data- and information-related areas, which is reasonably common – for example, information security tends to still sit within IT and your DPO will probably report into a governance or legal function – then you need to use all your stakeholder engagement skills and get everyone pulling together on this one. The policies will all interact with each other so you don't want your information security policies competing and disagreeing with your data retention ones. However, to achieve this union it needs to be in an area where one team's idea is firmly in place. The rest of the company do not need to know about how you and other data- and information-related professionals (and yes we are including anyone who looks after knowledge management as well in this) are collaborating on this. It is not their problem that there are different reporting lines or areas of responsibilities, as far as they are concerned data is data, it is a tool that is so fundamental to what they do that they probably don't realise the significance of it (yet). Why would you make this more difficult for them than it needs to be? For their sake, and as far as the rest of the business is concerned, you need to be one team.

Your watch words when it comes to policies are simple, consistent and relevant. As we have mentioned before, all too ofte we have seen policies or control documents that are written to rival a copy of *War and Peace*. Unless you are putting detailed instructions together on how to launch the space shuttle then we would seriously question if you have the right level of detail for what you are trying to do. Now it is not for us to say that you have too much detail but if the reader falls asleep before you get your point across then that might be a great indication.

When you are putting policies together try to keep the level of detail appropriate for what you want to convey. Remember you are aiming to help people understand what you want them to do, so don't make it complicated. If you keep it simple then it makes it easier for them to comply with the spirit of what you intend as well as what the instructions say. Remember, you can alternate between different levels that hit the right amount of detail for various parts of the company. It isn't necessary to shove everything into one document. Two or three pages please, not a novel – and put the important stuff at the beginning, don't make people search for it!

The last point about policies is the importance of making sure they are relevant, which ties implicitly to keeping them simple. If there is no need for the policy then why on earth do you have it? All you are doing is cluttering up what is possibly already a jungle of paperwork. Companies can be quite good at creating policies but maybe not quite as good at deleting old ones. Is it because it is easier to add another one rather than refine a similar one? And does everyone know that's what you have done? It is amazing what else we can find to do rather than tidy up. Before you try and update something or make it the right level of detail, you have to check that there is a need for it in the first place. Why waste your time or your company's time fixing something that shouldn't exist in the first place? It also makes it a great deal easier for everyone who has to work with those policies to understand what they are expected to do. It makes life far simpler and quicker if you have less to search through to find what you are looking for.

There are only two general reasons why people don't comply with policies, the first is that they deliberately decide not to and that needs investigating. Are they not complying because of malicious reasons? In that case there should be other policies within your organisation that you need to use to deal with that. However, you also need to check if it's because they genuinely believe that the policy is wrong and they think there is a better way. How you deal with that is up to you, but we would at least start with the premise that they are right and give them the benefit of the doubt before judging them in the same category as the people with malicious intent. If someone has deliberately not done something that is serious and should be treated as such.

The second reason is that they couldn't comply, they didn't understand, didn't know about them, didn't have the tools or the training to be able to in the first place. We will deal with the tools and training in Chapter 8 but this section is very much about the understanding and awareness of them. If someone can't comprehend what they are supposed to do because they are trying to decipher the dead sea scrolls, we really don't see how you can blame them when they don't get it right.

Putting together an upside-down policy tree (or whatever you want to call it that sounds better than that but looks similar) can help with the understanding of the policy estate (see Figure 6.3).

So, we have talked about why simple and why relevant, but why consistent? Well that's simple, it makes perfect sense to not confuse people. If you read something in one policy, you don't want it to be contradicted in another one. It's absolutely amazing how often this

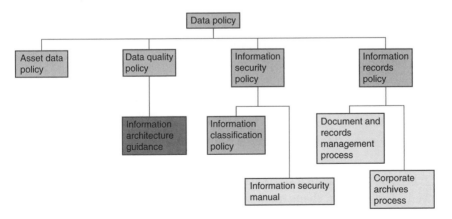

FIGURE 6.3 Policy tree

happens though. Due to the silos within organisations policies have often been written that unwittingly compete with each other, usually because the creators were unaware of what other departments were doing but occasionally politics might have played a part with one area disregarding what another has already created. Knowing what you are facing helps in dealing with it.

Looking at your policies, the high-level areas that you need to cover are:

- Policy for policies

- Classification

- Sharing

- Security

- Data protection

- Governance

- Analytics

- Data science

Again, there is no one perfect picture that we can give you to cover all the policies that you need to at each level, it depends on the type of company you are and how mature you are when it comes to policies. At least we have given you the big headlines that you need to address.

Getting this picture in place is really helpful as it gives you the image on the front of the jigsaw box – now you just have to find all the pieces and put them in the right place, simple right?

If it ain't broke, then don't fix it. There will be lots of policies that do need to be written, changed or tweaked so if something is working don't change it. Pick your battles carefully and don't start ones you don't need to – well, not yet, there will be time to sort them out later.

The best thing you can do is compare what you currently have with what you need. We like to create a variation that highlights what you already have that you can leave in place and use, what you need to create and what can be modified rather than thrown out. It's very visual and helps focus your mind.

It's worth a small aside to look at what goes into a policy or control document. At a basic level it needs a context; what are you trying to do and achieve, be really clear and specific about what it is you want the reader to do, what the policy covers and what it doesn't cover.

A bit like the message we talk about in the rest of the book, don't try and do everything at once. Prioritise which policies and at what level you need to go after first then second and so on. Look at your upside-down policy tree (Figure 6.3) and see where you have nothing (you might want to start there!) and look at where your inherent weaknesses are. Pick and choose the areas to focus on carefully around what will be used first and what will be most important for the company to follow.

Like most of the things that we talk about here, you need to get agreement to this from the relevant parties – anyone the policy impacts or (much more likely) their representative. There is absolutely no point in putting in place a policy that just imposes itself on part of the organisation, the battle that you will constantly fight is for the heart and minds of the people. Companies are so busy and full of competing demands that it's easy for attention to slip and for people to forget about the value of the data they are looking after.

Make sure that you use each of these meetings as salvos in that battle. Always be thinking about how you can be engaging people and changing them into data cheerleaders, or at least not data villains. You will immediately get people on the wrong side of this equation if they don't feel like they are important enough to be included from the beginning. If this is a horrendously large number, for the first one try and include as many people as possible and get them to sort out among themselves who they are happy to make these decisions. It is very likely that a lot more people will need to be consulted than those who will actually make the decisions that are right for the company. Lots of people will be using the policies so you need to get their view on the accuracy of what is happening in the policy and whether any changes can be implemented. They may not, however, have the full picture of the rest of the company and, without that, could take a more parochial view then you need. That does not limit the importance of their view though, their detailed opinion could hold some very important points – remember the devil is often in the detail.

Policies need to be factored into the framework. Once you get over the initial view of the policies and how to get them signed off the decision making needs to be an integral part of the framework. It's not just a case of getting them approved in the first place. You will be developing and changing lots of different policies so you don't want them all to be signed off at the same time. That would just simultaneously overwhelm and bore the company – neither is a state that you want to induce, it won't win you friends. Think about how you can get approval for a number that the organisation can cope with while still making a difference to the data-driven journey. It also helps when you go through the process to keep everything up to date so that you don't have to do it all at the same time. It's like when you start a new company and there are lots of online courses to get you up to speed. With that new job eagerness you power through them all and get them done while you have the time, which is great. But when the time comes for you to recertify and you are full on in the throes of your day-to-day job now, being faced with doing them all again quickly doesn't quite hold the same appeal as it did in your eager first few days. Spare your company the same thing and portion out the sign off and introduction of new and changed policies. It will make the adoption and engagement around them much easier too.

The final thing to make sure you don't forget is to make the policies accessible. You can have the best policies in the world but if no one knows where they are or can't get to them then they aren't going to have quite the impact you were looking for. Ideally, you will have, or establish, a system that puts all the policies in one place that everyone can get to, where the inter-relationships between them are demonstrated clearly and they are a simple 'one source of the truth'. If that isn't something that your budget can stretch to then as long as you can still have a collated set of polices, you know how they impact each other and everyone else can easily find what they need then that will do.

Chapter 7

Tools

T here is a bit of a clue in the title here, this chapter is all about the tools you need not only to be able to play the data game but play it well: the information architecture and the right metrics and technology. Let's equip your company with the top tools for looking after data properly and, while you are putting those in place, we'll help them get the best from what you already have. Don't wait for everything to be perfect, perfection is a disease when it comes to change. Make the most of now and plan for a better future.

Architecture

Rohit Singh, Head of Data Architecture and Data Strategy at Santander Global Corporate Bank

Information architecture plays critical role in data governance, data integration and automation within any organisation. It brings people and technology together. Also, information architecture looks into all aspects of the different data sources that provide structured, unstructured and other forms of data. Information architecture components such as classification, information lifecycle, metadata management, definition and naming standards and security models provide the foundational layer and consistent specification for any organisation.

Information architecture adds dimensions to data, that is, content, context and users. Any company can add value to their data asset if they know in what context

*data is consumed. Information flow is a vital component for any firm to cap-
ture Meta data around it to drive their business. When integrated with your
organisation's information platforms and tools like content management systems,
digital asset management systems, search engines, websites and portals, it will
improve content organisation, accessibility, reuse and findability, making it more
valuable.*

*In today's world organisations have diverse data structures within an organisation
(e.g., structure, semi-structure and un-structure) and data volume is increasing
day by day, so understanding the trust of association with every component of
information across all systems will be even more important. The success of data
analytics and insights will be heavily dependent on the architecture framework,
which enables information exchange and sharing. Information architecture helps
organisations to overcome challenges to gain access to the information they need
when they need it.*

Jacqueline Harrison, Head of Information Architecture at NFU Mutual

*Information architects, as individuals, may operate inside the Group IT domain
or belong to business data teams. At heart they are working to the same goal. They
work in both project and business as usual modes to make sure data and data
control aren't afterthoughts!*

*Within a project, they contribute significantly to the solutions architecture pro-
posals to make sure the data perspective is fully addressed. Too often I've seen
a business requirements document list out long streams of functional needs but
give no clear insight on the 'data and information' needs. The business process
cannot function without the right data! The architect plays a part in requirements
gathering – why do you need that data? To support what process? They make sure
we architect solutions for the benefit of the broader organisation and not create
more silos. They use data models to raise the questions nobody else thought about,
for example, can this asset be installed in multiple locations over its lifecycle? They
think about broader requirements – what document requirements and controls are
needed? Are there any data related reporting or regulatory obligations? They own*

or have a major role in creating migration strategies – what data do we need to move over to the new system? What options have we got? Where do we source it from? At the end of projects we need the information architects to 'harvest' back the valuable sets of project documentation that can too frequently get lost in project archives.

The information architects would typically act as custodians for the catalogues or inventory related to our data recording what data we have, where it is, who has accountability, how we define it, that is, managing the standardised glossaries, dictionaries and reporting catalogues we need to make available for others to exploit the data. They generally have a network of contacts so even if we haven't got a fully completed metadata set to work with, they will know who to contact and ask! They make 'data work' by enabling people to answer their data questions about today's world and help shape what's needed for maximising data value in tomorrow's world.

Data architecture is composed of models, policies, rules and standards that govern which data is collected, how it stored and arranged and, basically, how it uses this data. This is a fundamental aspect of how IT departments organise their systems around the business, it is all part of the overall architecture, as the pillar of an enterprise architecture. That's all well and good but it is very technology centric, it focuses heavily on how the data is processed and utilised in the different systems. That doesn't mean, however, that it isn't a useful discipline, it's crucial and working with your data architect, who is already in the company, will make a massive difference in getting your information architecture in place going forward. The process of working with enterprise architecture to tackle the different models is also very useful and we can use this as well, going from the conceptual to the logical and physical different layers of how to describe your data technology estate. It really helps to understand this. For anyone interested in this particular topic then ask your IT department about TOGAF (the Open Group Architecture Framework).

So, while data architecture is incredibly important, that skill will prob-ably sit within your technology department amongst the other architects (whether they are enterprise or system architects) who will focus on what the technology estate looks like currently and may look like in the future for the business. That is already taken care of.

What we need to focus on is the understanding your business has around the information architecture. We know that what we talk about can overlap with data architecture and may be viewed as part of the data architecture. From an architecture point of view, it's all part of understanding the same big entity but coming at it with different views. You can't look at the data in isolation, just like you can't look at the technology in isolation, at its heart it is just simply different views of the same thing. We also know that originally information architecture, as a discipline, was focused on the internet design side of things but it has grown to be something much more useful.

The difference between data architecture and information architecture for us is about focus. The data architecture is focused on how data is used in the systems and the model of where it sits within that technology. Infor-mation architecture is interested in how information flows throughout the whole organisation: it's interested in where the data comes from, why and how it transfers into information and where the ambition is to turn that into knowledge and wisdom. It seeks to understand what happens to the data on that journey. What are the key points that change it into a different state and who is accountable and responsible for the different groupings of data information as it moves along that journey?

When we are looking at what is driving the business value chain and decision making it is a view of the information. As you start to decom-pose the hierarchy, ultimately you get to the lowest level of data – this can then be mapped to initiating business processes to define lineage and reuse opportunities. A good enterprise architect will be looking at the full alignment to drive forward business change and will not be technology focused per se; however, they do sit within the IT departments usually and are driven in that direction so we think it's a good idea to have an information architect within your team.

The different architects within your organisation should have a very symbiotic relationship, they are all performing a useful role and by working together they can do a better one. We often have the information architects in our teams working alongside the enterprise architects because, really, they are all focused on the same thing.

Saying that the value sits within a single piece of data is like saying that a single block of Lego fuels a child's imagination. It is the full range of vibrancy, the combinations and the vivid colours that transform single pieces of Lego into a spacecraft that will launch you to a different universe or create a new form of life such as a flying pig. As we mentioned before, the value in the data is when you combine all the different sides in such a way that the sum of the whole becomes more than its composite parts, just like working with people. What we need to do with information architecture is understand the current state of what is happening with the data through to information within an organisation and what the ambition is. What causes those changes to happen and what can you do to speed that up? Going back to our Lego analogy, it is giving you the basic instructions and then keeping those instructions up to date when you morph and grow into something even more impressive.

Information architecture asks question such as whether or not the current state enables the organisation's ambitions on how it treats data. Does it underpin the data strategy and more importantly does it fuel the business strategy? Basically, without information architecture it's a bit like trying to look after a library the size of the Library of Alexandria in Egypt without any form of record-keeping, indexing or method of categorising the books. In fact, considering it in terms of a library is a really great way of thinking about your data, the books in the library don't move location based on whether you're looking at them from a particular point of view, you can search for a particular author or, if you are looking for a particular genre, the book in question will always be in the exact same location on the exact same shelf, what you have is a great way of utilising different ways of finding the book. When it comes to information architecture you are looking to replicate this kind of idea with your data, to create and understand the systems that help bring its value to life.

As with your strategy, a key area of focus is business alignment. It is important that the corporate information model reflects the organisation and can be recognised and related to by the business at all levels. A good way to achieve this is through alignment with the corporate value chain. What are the key things that the company does to deliver value to customers and shareholders? Understanding this allows you to link the core information entities and identify what information is critical to business success. Other equally important methods are identifying the information that drives key corporate decisions and performance reporting, for example, Balanced Scorecards, regulatory reporting, corporate governance reporting and so on.

The common information model, that you build, provides a canvas to create many views of the organisation and focus on where to start with your maturity. For example, aligning information domains to key business reporting with an overlay of quality allows you to pinpoint the more important information within a particular business context and hence to prioritise your efforts.

How much control do you really need over this data? At the highest level you should probably have somewhere between five and ten different data domains at its biggest conceptual stage. Break down all the dates you use into this structure and see if you can identify data that doesn't fit? If you can then you need to expand the remit of your data domains. Don't overthink this stage. In a lot of cases the biggest data sets follow patterns within the business, so look for those patterns. The really hard part of this exercise is making sure that you have people who are accountable for each of those data domains. This is important as it also ties in with your governance structure because those people across the highest level of the different data domains are accountable for making decisions for any piece of data within the organisation. They are therefore the people that you will want to be tackling your highest level of decision making when it comes to data. However, we spend more time tackling that within the governance section in Chapter 5.

Some parts of the data are set quite simply and readily within certain areas of the business; for instance, it's highly probable that any piece of finance data will sit within a single domain and very likely that your CFO will be accountable for that. That's brilliant for that type of data; however, it's not quite as easy when you're dealing with some of your more operational data. An example would be customer data, for instance – where specifically does your customer data sit? It probably moves through the lifecycle where it enters through some kind of potentially marketing or sales type activities, suitable operational parts of the business please don't necessarily lend itself to the same executive. That said you still have to have somebody accountable for the majority of the time it spends within the organisation. It is highly unlikely that any piece of data, if you want to get the full value from it, is under the sole responsibility of one person through its whole journey. Either it will sit within an area for the majority of the time or one area will derive the greatest value from it and hence be more interested in it. We can't give you an exact answer as to where as it really does depend on your company and who is willing to step up. By all means go after the low hanging fruit, as long as it's the right kind of fruit.

Looking at who is willing to take ownership is also very, very interesting when it comes to stakeholders. It can help you understand your stakeholder position; perhaps everyone buys into the idea of something being done but they just don't want to be the ones doing it. You also have a lot of busy people but if they find the time to agree that they are accountable then they are demonstrating active commitment.

Now you have those highest levels that's a great starting point, it's just not very likely that you'll make all the decisions at that level. Look at which ones need to be broken down into areas that are potentially more manageable for the organisation and again very importantly who is accountable for the data within that subset. That was a case of rinse and repeat. Don't try and go down to minute level of detail that will just tie you up in knots, what we need to do is get down to a level that the organisation understands and is comfortable with, with accountable owners you can have very sensible conversations about the detail if needed. In a lot of cases, if they are accountable for making decisions about changes to their data within their own silos, as long as they're ethically covered and they adhere to the right levels of governance, then it doesn't matter what they do with their data. It is when they interact with other silos that you have to have clear relationships in place. What's really important is when data has to can be combined with other data from other silos. This is where you start to get real value from it, but you need to have the right organisation in place to allow the right decisions to be made.

Once you have your big domains – or smaller level domains depending on what level you're working out – you can now start to look at how the data flows through that and at what points that data changes into information and beyond. Don't assume that data moving through to information, knowledge and insight is a linear journey – if only data worked in that way. Data can act like a living breathing organism and as such can take random jumps through your organisation, hence why information architecture is so important in order to get an overview of what it's doing and maybe help nudge it in a better direction for your company. The data is a perfect example of something working in a very agile way.

Disconnect your information architecture from your organisational structures as they aren't the same thing. Information is used across the organisation supporting many different processes aligned with multiple stakeholders. The conceptual information model will be a representation of corporate information and relevant to all business areas. Ideally the model will not change to reflect new business processes or significant organisational restructures as it should be a stand-alone artefact. This may not be the case in the event of introduction of new business models, requiring new information sets but that is all part of keeping your models up to date at a relevant level.

This is about the current state of the organisation, but does this match your data and business ambitions? The real art of information architecture is in your construction at the data domains that will make these ambitions real in the future and then work with the enterprise architects in your technology department to look at what tangible actions have to take place in order to drive through outcomes.

This all links to the information asset register as well. The definition of the conceptual information model, decomposed to the appropriate level, provides the foundation for the information asset register, a common representation of the corporate information and data along with definitions, ownership and so on, which means you build up that common language across your company.

Working on your data lineage will pay dividends, there are different ways of doing this and we have seen lots of people using charts, spread sheets and various diagrams. This can take a great deal of time, the human effort required to make a significant difference in this area, to get to the point where you have a reasonable amount of coverage, is considerable is considerable –. There are many tools now that can help you in this area specifically. We highly recommend that putting in a data lineage tool should be one of the first elements of the technology changes you make. It's much easier to look at the different stages of your data and also demonstrate to your stakeholders if you have a dynamic system through which to visually demonstrate what you are talking about.

Using a tool also provides a massive breadth of coverage very quickly compared to manual intervention. Having a full picture of your data lineage across your whole business gives you a powerful representation of an important view of your company. It's like looking at the lifeblood of your organisation, this is what keeps the machine of your business working. It keeps you working in a more 'real time' fashion so that you don't have another data governance problem to deal with to keep this all up to date.

Just having a tool though doesn't mean you still don't need to do the work. We aren't going to be abdicating our responsibility again. It gives you the ability to track your data flow through the company and it does give you some of the answers – but you still have to ask the right questions.

The tool doesn't complete your data asset owners for you as well, those people who are accountable for different areas of your data. Knowing your data areas and the lineage of those domains puts you in a great position but if you don't have accountable people who all agree they are accountable for those parts of the data and are prepared to work with the other accountable people who are in similar positions then all you have are pretty diagrams. It's never quite as black and white as that. You have data heroes who will pitch in but you should be trying to rely on the proper roads not the short-cuts that your data heroes carve out. The software can keep your data lineage up to date in a more simple fashion but it doesn't automatically keep your asset owners up to date. This is why regular reviews need to take place to check that this stays relevant. Factor this into your framework.

Why should you go to all this effort to get your data lineage right? The simplest answer we can give you is that you will be shocked by what you find if you do. Let's take the start of the journey as a starting point, the beginning feels like a great place to start usually. An example from an organisation we have worked with and completed this exercise with involved buying in data from external sources, like a lot of companies are doing now. In this particular case they were buying geographical survey data that was really useful to them from another company and they were getting a lot from it, so that is great right? Well in that circumstance then yes, it was great, they were buying in quite expensive data but they were getting value from it so that worked out well on balance. They were happy with that situation as they could see the value they were getting from the equation. They were happy until we pointed out that four different parts of their company were buying data from the same company, with four different carefully structured data sets, which were all relatively expensive to bring into the company. Of those four different sets, there was over 90% commonality between the data sets. By working with the different departments who were purchasing this data and making some small compromises we were able to reduce this external purchase down to one contract, which still satisfied each different area of the business. This was a significant saving for that company. To put it in perspective that saving was more than three times their entire investment into their data-driven business transformation. That's not a bad return on investment. Especially when that was one initiative and their programme brought them other dividends as well.

It wasn't the technology that made that happen but it did provide a way to enable the process start. So where is the obvious place to go from the beginning – the end of the line obviously. Where does your data end up? Where does it go? Does it turn into something even more useful? Does it grease your company's cogs? Or does it end up in dead ends? Is there a data graveyard sitting within your walls where data goes to slowly die?

We can't give you examples of everything you need to look for in this area as there are so many that it would be a book in its own right. The key areas to look at are where data ends up going nowhere and doing nothing (this is especially important if you are spending time and resources in order to put it into a place where it does nothing) and repetition of data (do you have 50 different versions of 'customer' – you might need that, we doubt it but you might, and what are these being used for?

Tailored reporting is the last area we suggest you also focus on (while this is not strictly information architecture it is useful and fits nicely with this section). If you are using an army of people to create reports for another army of people to use you need to ask yourself if that is the best use of your organisation's resources. Might it not be better to have the vast majority of your reporting standardised so it was easier to make sure you have a more limited version of the truth through your company with a smaller number of the tailored versions. Reports are one of those things that grow like weeds in companies, people create them and then never shut them down when they are done with them because they aren't important. People leave organisations and sometimes the need for the report goes with them but they don't take it, they leave it running to clog the arteries, leaving worker bees happily wasting time producing them because that's what they have always done. The other factor here is that, with the proliferation of end user tooling and on demand business needs, there may be no need for standardised reporting (but we would suggest your organisation needs a high degree of data maturity before you can fully capitalise on this ability). Information and data architecture allows you to define, catalogue and share the structures and associated definitions that enable integration and sharing of trusted data sets. This allows end users to develop the insights relevant to their daily operations rather than investing in standardised reporting that often does not meet the needs of the majority.

The real key to information architecture is, once you have this all laid out in front of you, what do you do with it? Now you have to build up a picture of the target state you want to get to that will really underpin and enable both your data strategy and business strategy. This will then be a key element in working with the process and technology parts of your company in order to help them understand what you are expecting from them. It's a visual representation of how you want the data to flow, so you speed its journey through to wisdom allowing you to work closely and simply with your IT department as you both head in the same direction.

Metrics

Heather Savory, Director General for Data Capability at Office for National Statistics and Co-Chair of UN Global Working Group for Big Data

Metrics are an essential part of assessing progress towards your goals by providing information about what is going on. Correctly scoped, measured and interpreted they help in the delivery of any strategy or operational service. They should provide insight into whether your resources are well employed and help you decide what to do more of and what to do less of.

Galileo is reputed to have said 'Measure what is measurable and make measurable what is not so'. Whether or not the attribution is correct this is sound advice.

Peter Drucker's famous quote 'What gets measured gets managed' is also helpful, but not always true. Many things which are managed are difficult to measure. It's important to understand what you are doing, and work out if your metrics are genuinely broad enough, including non-numerical measures as necessary, rather than measuring what is easy to measure to provide comforting, colourful dashboards.

Finally, one of the most challenging aspects of measurement, is to make sure you're not getting too familiar with your metrics – remembering that they are often an overhead for the organisation. Review them frequently for suitability, accuracy and utility and consider how often they should be modified – particularly if you are trying to drive innovation, growth and new behaviours

Be careful of your metrics here. You will measure something, whether it is your KPIs or your early warning indicators; just remember that what you measure drives behaviours. In one instance, we had a helpdesk reporting into us; there were reasonable customer care scores but nothing startling. One of the key measures driving the team's performance was how quickly they could close the first line call. So, naturally, the first line advisers weren't really trying to help the customer, their target was to pass them on as fast as possible, which created a load on the second line and didn't exactly make the customers happy either. After getting rid of that measure the customer care scores went through the roof. The first line advisers trained themselves to deal with calls they would have happily passed over under the old regime; they were happier, as they enjoyed the variety of their role more, the customers were happy because over 80% of the calls were now resolved within one phone call (which might go on for 30 minutes, but that was preferable to the possible three-day wait to get the second line team to solve the problem) and it had a knock-on effect on the second line support and beyond, who all had more time to deal with the more interesting problems. It saved money, as less was being sent out to suppliers to solve for us because we just didn't have the time to focus on them – it was a true win–win all round, achieved by thinking carefully about how we measured what we were doing.

The measures should be part of your assurance activities; even a Maserati needs a tune-up now and then just to make sure it is running optimally. As with all the processes you have put in place, weed the path, check the signposts are still in place and are clear and make sure that everyone still buys into the journey. So factor what happens as a result of the metrics. Don't measure something if you don't want to either track if something is possibly going to happen (early warning indicators) or if you want to drive a particular behaviour. Otherwise why bother? That would just be creating work for the sake of it.

Let's look at the two types of metrics, the early warning indicators and the measurement of your progress (or lack of progress)

Your early warning indicators should form a section on your risk but that might not be the only place that you track them. They will cover things like the number of data related issues you are tracking, how long these take to get resolved, possibly the audit actions relating to data and information, number or reviews that have been completed as a starting point. These are the really basic ones that you can track from the beginning, there are others that you will need to collect data on to see the pattern in the data before they can be of use to you (these depend on what you were already tracking) and these will depend on the changes you are putting in place. An important part of putting in place early warning indicators is the tolerance levels, you need to set the tolerance levels so that they are wide enough you don't jump at every little twitch because things change, but close enough that you get an EARLY warning, not 'an about to hit you' warning. Most people are nervous when setting the tolerances for the first time as they are not sure what good or bad looks like – honestly just put a stake in the ground and see what happens. There is no bad starting place, only a starting place. Then start to track it and see if the tolerances is giving you enough wiggle room but still doing their job. Some examples to look at putting in place over time are things like the average time to market new insights, usage insights, track reductions on time to make decisions or how long it takes to fix something that has gone wrong and definitely look at tracking data quality just be careful about how you define that.

Remember you might make it appear worse to make things better but you weren't tracking things before so you haven't actually made things worse, you've just shined a spot light on things that were already happening but were hidden. Or you are tracking them to a consistent standard which may not have been happening.

There is no point in having metrics, of either type, if they are not tracked and reviewed so build into your process how this happens and keep on top of it.

When it comes to the metrics that provide your measurements across the company and drive your behaviours, these you have to be careful with because they are a powerful motivator.

Ask yourself the key questions, what measurements you have in place to demonstrate progress, if you are starting from scratch to put this in place there are some areas you can look at to make sure you get the right coverage across your business. Think in terms of the V's of data; Volume, Velocity, variety, variability, veracity, visualisation which, by getting these right which your metrics will measure your progress against will demonstrate your data value.

Volume – you can track how much data you have but remember that bigger doesn't always mean better so be careful that you don't fall into that trap. You might want to track a reduction in your data or look at the data you store about your customers.

Velocity – this is about the speed that the data is accessible, for your organisation this doesn't mean real time access for everything as you might not need that and don't waste time or effort doing things you don't need to do. You could have measurements about how much reuse you have in your processes as people can't find the data they need fast enough.

Variety – do you have the breadth of data that you need to use it properly or do you have a wide variety of data that you don't use?

Variability – is your data consistent? This is the area that looks at the 'one source of the truth' what metrics do you have in place to check that this is working and that you don't have people who think they are using the same item of data but are a million miles apart.

Veracity – is your data accurate. We want to make sure that when it comes to your data you don't fall into the trap of garbage in, gospel out. Garbage doesn't magically sort itself out without external intervention.

Visualisation – how is your data being presented, gone should be the days of lines and lines of excel spreadsheets being the window to your data, what visualisation are you using, how and why?

All of these things will demonstrate the value of your data to the business as you track progress. And demonstrate your capability.

Look at each area and see if the metrics around your data will help you demonstrate progress towards your data strategy or drive the right behaviours to win the hearts and minds battle. Are there other metrics which will demonstrate the link between the better use of data and your business outcomes?

Once you have the right metrics in place, has your process for reviewing them been clear and understood and constantly followed? Then can you demonstrate that actions take place as a result of tracking the metrics, if not don't bother tracking them. If actions do take place, are you getting the outcomes that you want as a result? Are the behavioural changes positive as opposed to negative or static? If not then change the metrics, don't set things in stone. Listen to the feedback and change accordingly. Make sure you keep your focus on the strategy you are aiming for and monitor your steps towards it.

Technology

Alban Gerome, VP of Data, Barclays

Peter Drucker famously said, "if you can't measure it, you can't improve it." This quote and other factors have sparked a data storage frenzy, huge advances in data storage technology where cost, size and power requirements for that storage keeps dropping year after year. In 1990 the human Genome project started and finished two years earlier than expected, silencing sceptics claims that the human DNA simply contains way too much data. In 2017, the ecole polytechnique de Lausanne has stored for the first time "tutu" by Miles Davis and "smoke on the water" by deep purple on DNA. Steve Jobs wanted to put 1000 songs in your pocket. The whole internet stored using DNA as of 2017 would fit… Inside a shoebox. When organisations recognise the value of data, data that is getting easier and cheaper to store every day, many decide to store as much as possible, for as long as the regulators allow them to. This data capture practices bring challenges regarding how to extract the value from the data. Not all data is equally valuable the vast majority of that date has no value whatsoever. The advances in data storage must match similar advances in separating the signal from the noise, the actionable data that will help your organisation make data inform decisions without the distraction of the "so what" data.

Why should companies be data informed rather than data driven? I believe that the data technology vendors are the primary advocates for data driven approach. This approach can easily alienate swathes of senior managers with decades of experience who suddenly feel like their decades of business knowledge is irrelevant, and they are now under constant conservatorship. People want to portray themselves as making rational decisions from hard facts. Neuroscience has demonstrated that emotions play a much more significant role in decision making than we would like to admit sometimes. Indeed, most people make decisions on an emotional level and then use the hard facts and the technology provided data only a posterior so they can defend their choices. Erik Brynjolfsson and Andrew MacAfree revealed in an HBR article that many organisations are only really date justified, cherry picking data that justifies priory held beliefs. New technology is also prone to hype, reaching unrealistic expectations fast which are hard to manage once confronted with reality. As unfortunate as such practices may look, they provide us with the valuable lesson that we should be seeking a compromise between emotion and data, technology and experience, a more data informed approach.

Abboud George Ghanem, RVP, Middle East and Africa at Alteryx

Technology as defined in the Oxford dictionary is "the application of scientific knowledge for practical purposes" is a key pillar in the CDO's life in the organisation. What interests me the most is the emphasis on application, knowledge, and the practical element.

I see organisations from all sizes and sectors spending millions of dollars on big data platforms from databases to data warehouses to data lakes to cloud services and so on. These organisations often miss out on an important piece, delivering fast knowledge that can be applied in a strategic business context to the business user or decision maker.

Why is it difficult to do so? Historically, extracting data from multiple sources and having it ready for analysis required different teams with different IT skillsets to do so. As a result, the business users gave up on large business intelligence projects that are time consuming and difficult to learn and use. Hence, excel has become the most dominant analytical tool in the business today.

To apply the knowledge from data, organisations need to bring knowledge and insights production closer to the business, where deeper and adhoc questions are being asked. Often these questions are answered with gut feel and "how we do it here" approach – because it takes too much time to get the answers from specialist teams or IT. With the availability of easy to use platforms like Alteryx, organisations can upskill their smartest people in a short period of time to deliver newer and faster insights to the business without the need to code.

What is the point from having technology investments that don't address business questions and assist leaders in their daily strategies? The application of knowledge is key for the CDO success by delivering a programme to drive a data enabled culture that impacts the business's top line and bottom line. the most successful programmes I've seen are the ones that make the business users feel empowered to deliver more value from data at a fraction of the current time, and with continuous enablement and learning – all with senior executive support.

Alban Gerome usefully highlights the connection between emotion and the technology and we love his take on being data informed. He gives us the reason why the technology is important but is not the primary driver behind the change we are advocating across businesses. Talking about all data not being created equal is also a theme we advocate, don't treat your Crown Jewels in the same way as you 'thrown away' data. That said, technology plays a strong part in underpinning the data strategy and working with the technology means you can cope with the swathes of data your company throws at your technology.

We have a tremendous amount of sympathy with any CIO dealing with the changing estate of demands from their data. We have touched upon having to deal with legacy in other sections, looking at things like legacy attitude, legacy processes, etc. but few of them are as hardwired as the legacy IT estate. IT has probably invested millions over the years in the systems that the company are using and they are heavily invested in it. You will know where you are and the situation you are facing from the maturity assessment so there are different parts that you need to break this element down into to ensure your technology estate underpins your data ambitions.

Out of everything you first thing you need to ensure you have right is the relationship with your IT department. What level of control do you have over the data related IT estate? We aren't suggesting that you need to turn into Ming the Merciless and become a dictator lording it over your IT department. There are many really good, skilled people who are experienced working with your technology estate. Let them bring their skills to the party and you bring what you are a specialist in so that you get the best from everyone. You do however need to be vested in the IT direction when it comes to the data technology.

Getting the roles and responsibilities clear is especially important in this area as you need to work efficiently with the CIO or CTO. Most CIOs will be really clear that one area they wish they could have more help on is with the business being clear about the outcomes they are aiming for. A mistake we have seen time and time again when the business and IT work together is people from the business coming with the piece of technology they want already fixed in their heads because they have had a really enthusiastic sales person waxing lyrically about how fantastic their product is and the business person is completely convinced that this one product will solve every problem they can even conceive. Let to just one isolated product it may be able to make an impact but this is where the legacy stuff comes back to bite you. There is very little that works properly in isolation, data comes from somewhere, it has to change and become information (any beyond) and go somewhere – otherwise what are you doing with it? It is much more efficient and effective to do what you do well, understand what you are trying to do with the data, what outcomes are you looking for and what are potential steps on that journey? Basically what is your strategy, we have already covered that in depth so we don't need to go over that again but the point to reiterate is to make sure your IT team are heavily involved in developing your data strategy and you should be as heavily involved in developing the data part of the IT strategy. So that longer term the two can converge to a comfortable place.

The problem for the IT department is that they aren't just dealing with individual jigsaw pieces; they have to worry about the whole puzzle. You have the luxury of only worrying about the data technology landscape and that is complicated enough, they have to worry about the interactions of the whole landscape. There is no point in your finance systems being perfect if they make your asset maintenance unmanageable. So often they have to make compromises that keep the whole machine working that means a few squeaks can be heard across the board. At least acknowledging this and working with IT rather than against them will make your working relationship easier.

All of this is great for the longer term but what about the short and medium term. There is no use pointing blame at IT that you can't deliver on your data vision just because it will take time to change the direction of your IT legacy. It really helps if you have a good working relationship with IT so send time helping them understand how much you value their help and how you need flexibility to try new things to help data flow better and move faster through to wisdom.

As with everything it's necessary to prioritise what changes you are looking to make, pick some key ones and focus your effort then look at the impact. Does your data governance systems desperately need an overhaul or is would your data visualisation need a spruce up? Do you accept that data science takes a slight back seat while you improve the quality? Obviously we aren't suggesting that you only do one thing, it's just worth looking at what impact your changes are having before you overwhelm the organisation. Remember your change management principles and listen to the heartbeat of your company.

When we talk about minimum viable products, this is the area that will benefit the most from that view. There is a need to experiment with technology and see if it works for the company, don't invest heavily on the perfect solution – remember the phrase 'perfection is a disease'. There are some really great products in the data space to look after your data governance element or give your data citizens more power to their analytic capabilities. None of which you have to fully commit to without having a play first and seeing how they fit.

Try and pick a real problem that the company is facing (by now you should have a fair idea of some areas to go after) and pick a product that can help you with it. Don't tackle abstract issues when you have enough real ones to deal with. Just don't play with the suppliers. If there is a real potential then ask for their help and communicate with them honestly, if it doesn't work that is fine but if you never meant to use them and just wanted some 'free consultancy' that poses ethical dilemmas. We have a really naughty story about one company who used an analytical company to solve a problem for them with the clear expectation of work to follow. The work was presented, a photo of the solution was taken and everything seemed on track. What followed next is the bit we advise not to do. The solution was implemented and all of a sudden the communication was stopped with no explanation.

So you have your real world problems and a possible solution, get the right people involved both from the supplier side, the data side and the rest of the business side. Set the team up like your 'tiger team' (see people section for an explanation of that term) and with the right focus you can build amazing things.

Always have an idea on what will come after, we want to minimise any 'regret spend' as much as possible. Why waste money? If it is possible to solve a problem, how do you build on that solution rather than throwing it away and starting on a much larger one? We like to talk about minimal viable products rather than proof of concepts as you tend to throw away a proof of concept as you need to scale it up to tackle the real problems. Why do we do that? Just because you are trying to prove something works (or doesn't work – that's equally as valuable, just look at the original light bulb development) shouldn't mean that you have to be wasteful with your resources. There are loads of tools out there that will work with you to solve a small problem within your company in order to look at if that is a viable proposition to properly scale up that you don't need to through away. Think continuing to build, augment and add on rather than having to start again. We are essentially quite focused on not wasting money, time or resources all of which are limited, we don't want to do anything again if we don't have to. By focusing on the short term and just the short term you are missing a trick, you can focus on the short term but keep an eye on the future.

As much as dealing with legacy anything can be a problem it can also be an opportunity. Legacy IT can be a bit traumatic to change at times however there are ways of putting nice shiny wraps about older systems that mean you can make changes quickly and then work to update the back end parts in a time that won't terrify your IT director. Think of a magnificent swan dance. As far as your company is concerned you look serene, poised even, making great progress. They don't always need to see all the hard work that is going on under the surface. Sometimes they do, so they value the hard work that your team and the IT team are putting in (and this will be a team effort so make sure credit is given where credit is due) but often it is enough to know that progress is being made at a pace that the company can deal with.

Utilise what is already coming and what is already underway if this is possible. It departments have to make plans to get their budgets years ahead in some cases (this puts them in really tough positions as times when you look at the pace of change of technology, we worked with one organisation who had to set five year budgets because they dealt with national infrastructure and that was their funding cycle only the whole iPad and tablet development broke during one of their funding cycles and they couldn't have predicted that. Needless to say a lot of the senior management wanted to take advantage of this technical change however they simply didn't have the budget so in order to make it happen other things got cut). When you go in to work with the IT team, we are sure they will be thinking about data but in our experience they are more focused on data warehouses, data lakes and archiving rather than data lineage and visualisation tools. So what you are introducing is a need to things that may feel like it is blindsiding the IT department. Work with them. Work with what you already have, you can make changes by repurposing existing software. It never ceases to amaze me how much software big firms have that they don't use or licences that they have over bought; re purpose where you can.

Working with the IT department also helps by establishing those good working relationships you also get a greater level of input into what the IT strategy looks like going forward, giving your information architect an in with the enterprise architects to fundamentally shift to a data enable organisation taking real value from its data assets. It's a great place to look at what the technical future will look like but prepared to take steps on what can feel like a long journey.

Chapter 8

Overall Change Management

'The secret of change is to focus all of your energy not on fighting the old, but building the new.'

Socrates

As you can see, there is a fair amount to do to move forward in each element of the transformation process. Remember what we said at the beginning about balancing this across your organisation. Listen to the feedback and push, but push in a constructive manner than pulls people with you as well. Most importantly, as Socrates told us thousands of years ago, you need to have something worth moving towards. If Socrates, who is widely considered to be one of the founders of western philosophy, said it who are we to argue. You would also think that we might have learned to do change better in that time period but hey ho.

As a slight aside, he also said 'the beginning of wisdom is a definition of terms (data dictionaries), know thyself (maturity assessment) let him that would move the world, first move himself (lead by example).' We think he might have been one of the first data leaders too.

We often talk about your change journey as the internal data revolution, you want to win the hearts and minds battle throughout your company and turn everyone into data cheerleaders. However, that doesn't happen by itself, there are lots of things you can do to smooth the organisation's transition into a data-enabled company. Change is hard, if it wasn't we would be better at it, especially since we have been talking about it for a few thousand years. It's called a comfort zone for a reason – it's comfortable. It feels safe so people will fight to stay in the comfort of doing what they have always done rather than risk the new. That is, unless the burning platform beneath their feet is more uncomfortable or the change is so enticing that the fear of the change is less than the fear of staying static (or both).

If you address a group of people and ask them who wants to see change happen you will have a sea of hands demonstrating what a great idea they think that is. Now ask the same group who wants to change and you won't see quite as enthusiastic a response. It really is human nature to resist change and assume that a more established process is better. Studies have proved that people have a very reliable and tangible preference for things that have been around longer. For instance, people would have more favourable attitudes to acupuncture, for example, if they had been told it had been around for over 2,000 years than if they believed it had only been practiced for 200 years. There is something in us that equates how long something has been in use with how good it is. This is a rational approach and has probably saved lives somewhere down the millennia. If a way of working has stood the test of time then it is probably superior in some regards. So, it isn't just that people fear change, which we also think is a factor, but that they have a voice in their heads telling them that if they have been doing something one way for longer, then that way is better.

However, this isn't the most accurate indicator of how good something is for our companies; it has in fact led to inertia and resistance to change. It is an unconscious bias that the vast majority of us hold and you need to understand this, and how uncomfortable you are going to make the individuals within the company, before you will make things better. Just walking in and assuming that what you are selling is such a no-brainer that they would have to be stupid not to jump on the bandwagon isn't going to help you win friends and influence people.

Take time to understand what position they are coming from. What is the status quo and what are they fighting to preserve? It will help you articulate the 'whys' behind 'what you are doing. Motivation is key to making any kind of change happen. Another wonderful thing about humans is that once we understand why we are being asked to do something different we are more prepared to look objectively at the change.

No one likes being forced into a corner and that is what it can feel like if you don't understand why something is happening and what the benefits of it (to yourself and the company) are.

Someone has to lead the change – that focal point that we talked about when we discussed the organisation element. During the programme of change that someone might be one person, but as the process moves to business as usual that someone might be the data leader. Words aren't enough here, everyone has to lead by example. Let's go back to paraphrasing Socrates again: 'Be the change you want to see'. There doesn't have to be just one leader, share the load. Why should you do all the work when you don't have to? Get people involved in the journey, this is literally a case of the more the merrier.

You will be clear about what you want to happen by creating your vision and then developing a strategy to execute that vision, which is one of the first things you will have looked at. This underpins your data nirvana, gives you the basis of the compelling story that you are selling to the company to convince them to come on this exciting journey with you. It also develops the clarity that you need to keep everyone focused on the same goal. You know what 'good' looks like, so use that, be evangelical about this picture of the future you are painting. Remember to also create a sense of urgency to avoid the 'why change now, can't it wait for a bit?' and 'there is always something more pressing, so let's put this to the back of the drawer' responses. Use both a risk-based and a value-based approach here, blending the two to focus on the area that means the most to your company; identify the threats and demonstrate the value you have to gain. We mean REALLY focus on the value: beginning with the risk is a starting point but move through that phase as quickly as possible and get invested in the value because that really excites people.

There is no point in doing any of this work if you don't tell anyone about it – so how do you do that? This is where your engagement piece comes in. No matter what you do you haven't engaged enough, don't just do communications (mass emails aren't the best tool for this) do engagement, listen and modify to better engage. The method that we have always preferred is using advocates: people listen to people they know and trust, and they are more likely to modify behaviour if someone they respect gives them good reason to do so. No matter how good you are at relationship building, unless the whole company is less than ten people you won't know everyone in it, hence the advocates. An advocate is someone who shares your vision, believes in the future you are creating and wants to work towards that future. They don't always start off being quite so passionate; on many occasions we have worked with advocates who have 'been volunteered'. While it would be nice to have committed, passionate supporters from day one, you are more likely to have sceptics who prefer the current mode of operation. By being clear, concise and committed to your vision, listening to their concerns and addressing them while promoting the positives, you can turn these sceptics into your biggest supporters. Now, if you have chosen to go through the matrix approach to your organisation you have data champions, which means you have an army of advocates. It you have a more centralised approach then you still need to find your advocates to help the change.

Don't assume that everyone learns, communicates or engages in the same way; people absorb information differently. While you need to be clear, consistent and tenacious about your vision, you need to constantly modify how you deliver that to the advocates in order to get the best from them. Listen to the clues they give you about what makes sense to them and work with those. Use e-mail if that helps but don't make it your be-all and end-all; try everything in your repertoire and see what works best, then do more of that. Once you start putting your message out there have a constant process of refinement and redeployment going on and then repeat. A lovely virtuous circle.

Remove the obstacles to help the advocates on their journey, whether those obstacles are poor tools or processes, training gaps or general incorrect perceptions. Find out what is stopping them or making their life difficult and deal with it. Make the new path easier than returning to the old one (use the shared pain: it can be a great bonding exercise). While we will never cease to be amazed by the resourcefulness and creativity of human beings in keeping the status quo in place, when constantly presented with an easier journey they do start to follow the path of least resistance. Make sure you take time to celebrate those successes – positively reinforce the right behaviour!

Lastly, don't expect any kind of meaningful culture change to take place overnight; you are looking at a journey that is measured in years not months, so to keep the interest up you are going to have to demonstrate that the path is the right one and that it's worth staying the course. Here you can use your immediate business value deliverables. Nothing makes people sit up and take notice as much as when you actually demonstrate a positive outcome. It makes them think that if you were able to achieve that in the short term then they could buy into the possibility of what they can achieve in the long term, building on these successes. Just don't rest on your laurels. Take time to celebrate those successes, promoting the art of the possible, demonstrate that the path to data nirvana is littered with great outcomes and benefits.

Chapter 9

Running a Business in the New Data-Driven World: Arriving at the Destination

W e are now going to introduce the concept of the dynamic data-driven business transformation (DDDBT). This doesn't really flow off the tongue and D3BT sounds a bit like an extra from *Star Wars*, so we call it D3.

This chapter is a bit of a contradiction to everything that we have discussed in the previous chapters. Up to this point we have been talking about 'data-driven business transformation' – positioning this as a process that has a start, middle and end. We haven't deliberately misled you but there is more to ending than just, well, ending. This chapter addresses that 'end point'. Many organisations are currently faced with the 'start' of a data-driven transformation and the reasons why they need to address this. They are identifying processes that they need to go through to be 'ready' to undertake a data-driven business transformation, which we have discussed at some length in the book already. We have also discussed how the data-driven transformation can be delivered.

Being contradictory, this chapter proposes that the true data-driven business transformation doesn't actually have a start, middle and end. It certainly has a start and a middle BUT no end. You may reach the desired 'end' state, the vision that you have communicated to the rest of the organisation, but now it is time to realise that true data-driven businesses don't stop, their transformation has no fixed end point but rather faces a continuous process. Perhaps unlike digital transformations or technology transformations, which deliver a platform or a piece of technology and thus as if by magic deliver the transformation, true data-driven transformation is a continuous process as new data, or new data morphologies, can be collected, stored, processed in new and different ways, analysed in ever more detail at ever more speed and harnessing the power of artificial intelligence and machine learning. Indeed, our ambition continues to grow and evolve at the pace of or faster than our delivery capability. How could there ever be an end state? The present end state is limited by our current imagination.

So, in Chapter 2, when we made it clear that when you envision the data-driven end state and communicate that to the business to gain their buy in, we meant that the end state that you picture is only a point, a state in time, and when you reach it, or even as you reach it, it will be possible to imagine more powerful end states. As the transformation moves forward, what may have seemed unimaginable or just too plain difficult will be possible and realities and further new horizons will appear.

This is a very important frame of mind to embed in your company as part of the data-driven business transformation. It doesn't stop, it continues and becomes the way the business operates, grows and moves forward. It becomes the way the business responds to changes in the market, or even (and this is the most successful) creates changes in the market, innovates and disrupts. Just think about Uber, they didn't respond to changes in the market, but instead used data to change the market and they caused a disruption.

In very simple terms, the first action to take once the end state has been reached is to create a culture in the business that fosters, supports and focuses on further data-driven transformation. We need to create a culture of D3. A culture that sets the corporate state of mind into continuous improvement, innovation and transformation. However, to be really successful in achieving this shift of culture, it should be part of the original transformation journey, the original expectation and the original communications and vision. The data-driven business transformation will fundamentally change the way that the organisation does business, how decisions are made, how the market is addressed, the products and the operations, it will also radically reset how the business grows and develops. This change to the company's growth and development may be more seismic than the more obvious data-driven transformation itself. It could be more painful and more challenging, and a harder battlefield in which to win the hearts and minds of the whole organisation to create the dynamic transformation. The approach may be ripping up the established methodologies, approaches and dogma around change delivery within an organisation. Let's look at each of the moving parts in more detail, going back to the classic people, process and technology triangle referred to in Chapter 1.

The people

The people in any organisation are in many ways the most important element. The people have the power to embrace and deliver the transformation. They have the ability to engage with the new ways of working and make them successful. Equally the people have the power to frustrate, delay or even stop transformation and change. This may be en masse across the company, it could even be en masse across society, if the whole workforce perceive a threat in the transformation proposed and do not trust or believe in the changes that they face. Or it could be individuals, or small groups, who don't engage with the transformation and act as a barrier to change.

> *'Of all the things I've done, the most vital is*
> *coordinating the talents of those who work for us*
> *and pointing them towards a certain goal.'*
> Walt Disney

We have talked about the importance of winning the hearts and minds of the stakeholders and the whole organisation when preparing and delivering a data-driven transformation. It cannot be repeated often enough that people will feel threatened by changes and often will not understand some of the complexity of a 'data-driven' world, or they may simply have been fed many misconceptions about what this may look by both media and fiction.

*'Data (or Digital) Transformation often mistaken as an
IT project rather than an everlasting state
underpinned by cultural change.'*
Rob Howes Collibra 2018

Equally as important as the early communications strategy to win the hearts and minds of the people is setting up an enduring framework or methodology to maintain the communication and the engagement of the people in the triangle. The people in the organisation may well have to adopt new ways of working, learn new skill sets (such as data literacy) and be forward looking rather than happy with the status quo. To achieve this would either result in or require a highly engaged workforce. It is a kind of chicken and egg conundrum. But either way, whichever comes first, the outcomes and benefits of having a highly engaged workforce are recognised – refer to Gallup Q12. Therefore, the highly engaged workforce is a highly beneficial side product of creating the D3.

What is the mindset of D3?

- Never satisfied

- Always questioning

- Always constructively challenging

- Always believing that processes and decision making can be better

- Always believing that the customer can be given better service, better products

- Fast moving

- Capable of pivoting

- Achieving results on minimal spend

- Celebrating the successes of the journey

These characteristics resonate like those of a 'start-up':

> '*A **startup culture** is a workplace environment that values creative problem solving, open communication and a flat hierarchy. Because new businesses must adapt quickly to internal and external market pressures in order to survive, a startup culture also promotes business agility and adaptability as being key virtues.*
>
> *The workplace values supported by startup cultures are increasingly resonating with business leaders at large organizations. As the pace of business speeds up, quickened in part by advances in technology, large companies are realizing that they could benefit greatly by putting more emphasis on startup culture values, including the value of the individual.'*
>
> Margaret Rouse (June 2014) Search CIO

Margaret Rouse stresses the importance of the individual, collectively the people.

> In many ways we are proposing a Kaizen approach within the 'people' element of D3. Kaizen is defined by Rouse. '*Kaizen is an approach to creating continuous improvement based on the idea that small, ongoing positive changes can reap major improvements. Typically, it is based on cooperation and commitment and stands in contrast to approaches that use radical changes or top–down edicts to achieve transformation.'*
>
> Margaret Rouse (July 2018) TechTarget (Search ERP)

The ten principles of Kaizen (the Toyota Way) do apply to D3. Because executing Kaizen requires enabling the right mindset throughout the company. The ten principles that address the Kaizen mindset are commonly referenced as core to the philosophy:

1. Let go of assumptions.

2. Be proactive about solving problems.

3. Don't accept the status quo.

4. Let go of perfectionism and take an attitude of iterative, adaptive change.

5. Look for solutions as you find mistakes.

6. Create an environment in which everyone feels empowered to contribute.

7. Don't accept the obvious issue; instead, ask 'why' five times to get to the root cause.

8. Cull information and opinions from multiple people.

9. Use creativity to find low-cost, small improvements.

10. Never stop improving.

The point about Kaizen is that we feel it translates well into the people part of the focus of D3. Kaizen rests on the principle of 'respect for people'. This book isn't aimed at reinventing the wheel, great ideas should be shared and applied. So, to truly land a D3 an organisation needs to restructure and respect their people along the lines of the Kaizen methodology, behave like a start-up and develop their own version of this 'culture'. However, beware that this has to be embraced, embedded and done properly. A sham or merely paying lip service to these principles will lead to the transformation failing and further transformations being viewed with suspicion.

Tendayi Viki, in an article in *Forbes* (23 September 2018), identified three human barriers to transformation. Though Viki was actually discussing 'digital' transformation, these barriers also apply to a data-driven transformation, and perhaps even more so. Viki identified 'inertia', or the tendency for people to do nothing or remain unchanged. This is something we all experience and have to be mindful of. Most organisations have been operating reasonably well, so why is there a need to change? The people in an organisation, for this reason, may be inert and not understand or even be able to identify the need for transformation. Perhaps the only time when the need for transformation becomes apparent is at a crisis point, when the business is failing either commercially or in regulation. It might be losing market share to competitors who have transformed or are transforming, or suffering under the threat of disruption. To overcome the inertia in the people, the leaders must lead. They have to paint a compelling vision of the benefits of transformation.

The second barrier that Viki identified was 'doubt'. This is when the leaders have painted the vision of the future state and the benefits are understood by the people, but the problem is that the people 'doubt' that the transformation can be delivered in their organisation. This may be a 'positive doubt' along the lines of 'this is big challenge, a big step to take, will we manage to pull it off?' Or it may be a 'negative doubt' along the lines of 'there is no way that our organisation can do this, we've tried this sort of thing before and its failed, so what's the point?'

The final human barrier to change identified by Viki is 'cynicism'. This is probably the one that most readily leaps to mind. This is characterised by the satisfaction that some people gain when a transformation plan or change delivery programme in an organisation goes wrong or fails to deliver. All transformation will need to 'pivot', and this is exactly the point that we are making in this chapter with D3. So, the 'cynics' can be destructive – if not dangerous – for the success of the transformation and even the survival of the organisation.

People who experience the first two types of human reaction to transformation can be persuaded by good leadership, using some of the steps proposed earlier in this book. That leadership will need to be constant, creative and inclusive. Those that fall into the type of 'cynic' need to be addressed head-on, they are either with the transformation or against the transformation.

In summary, Viki identifies the importance of the leaders, who are of course part of the 'people' element of the triangle. The leaders, and the quality, ability and resilience of those leaders, are vitality important to successfully drive the transformation. One of the core leadership skills is the power to create and deliver a narrative that will win the hearts and minds of the people, and that is powerful, persuasive and honest enough to counter the negative forces of the cynics. Perhaps the leadership team also has to be strong enough to 'face down' the cynics one way or another. It is also important that you don't underestimate the resilience required by the leaders to deliver a data-enabled transformation and sustain it. There will be many distractions – both internally and externally – people will come and go, resource bases will change, so resilience will be required.

We will mention the film *Money Ball* later in this chapter. In this film the role played by Brad Pit has to battle the cynics in his backroom staff. In fact, *Money Ball* demonstrates all three of Viki's barriers very well.

In a *Forbes* article Daniel Newman (26 November 2018) noted that company culture is the biggest barrier to transformation. It is also interesting that the importance of people to the success of delivering a data transformation was a recurring theme at DataTalks London 2018, a Carruthers and Jackson event.

Process

The second element of the triangle is the processes. What we will focus on here are the processes for change or transformation. To truly embed and achieve D3 the change processes in an organisation need to change.

Let's look back to Chapter 1, where we talked about the differences between digital and technology-driven transformations and data-driven transformations. One of the major differences that we didn't discuss at that stage is that of delivery methodologies. Technology and digital transformations, in many cases, use language like 'projects' and 'programmes'. We are sure that you recognise these transformations. To get away from the odour of the two P words, some even adopt the word transformation, when what they actually mean is one of the Ps. It is their delivery methodology that gives them away. The methodologies use terms like 'requirements', 'project manager', 'programme portfolio', 'waterfall', 'deliveries', 'deliverables' and 'user acceptance testing'. They will often work to fixed budgets and fixed project delivery timelines and 'gateways' – which, ironically, they almost never meet or deliver against.

Data-driven transformations are more agile, faster paced, more business and outcome focused. We don't want to get embroiled in a debate over the relative merits of different delivery methodologies, but we are clear that to achieve D3 an organisation needs to be less 'project driven' and more driven by continuous improvement and agility to respond to business needs and outcomes.

If the Kaizen approach is accepted as a good model for achieving the people element of D3 then by extension the concepts of continuous improvement, small but significant deliveries, become understandable. Change has to be at the heart of the business, at the forefront so that it can be developed close to reality and delivered into operations quickly. Change must not be concealed in projects and programmes, hidden away from the business, delivered top–down and at the end of a programme.

Change processes need to be reformed as transformation processes. Change implies moving from one state to another. Perhaps a good example to explain this is 'changing' the pound sterling in your wallet to the US dollar, a change from A to B. Transformation, on the other hand, should imply a more dynamic state of constant change from A to B to C and so on.

If the 'change' processes in a business are going to become transformation processes, then skill sets, responsibilities and accountabilities will change around the process. This in itself will need to be managed and effected.

In reality, an organisation may run a hybrid environment for change. Things like application refresh, or network upgrade – things that are truly 'technology' – may continue to be delivered through a more traditional 'project' methodology, while the data-driven business transformation is delivered in a more continuous D3 methodology. Whatever the approach companies must set themselves up to have a delivery methodology that promotes rapid and agile delivery if they wish the data-driven transformation to be embedded, enduring and deliver business change and outcomes.

It is worth briefly discussing that data-driven business transformation should not be confused with research and development (R&D), which is a very different function and sits elsewhere in an organisation. R&D in the data environment may include such things as developing new data science platforms or new algorithms. It is the insights in the data that drive the business transformation, by enabling better decision making and decision-making processes. R&D may also sit closer to the operations and manufacturing functions and the size and importance of an R&D department will vary depending on the vertical. The point we wish to draw out here is that there should not be any confusion between R&D and data-driven transformation.

A final consideration for new approaches to process is the role of DevOps software methodology, which has a vital part to play in D3.

*'**DevOps** is the combination of cultural philosophies, practices, and tools that increases an organization's ability to deliver applications and services at high velocity: evolving and improving products at a faster pace than organizations using traditional software development and infrastructure management processes.'*

Amazon Web Services

https://aws.amazon.com/devops/what-is-devops/

This approach is vital to embedding a successful data-driven transformation. We would even suggest that there is a complimentary role that sits alongside the DevOps role and that is what is being identified as a new capability or function, 'DataOps'. Even though some organisations are, or are close to, deploying this function it has not been fully explored or defined. We have been working closely with Tariq Bhatti, from Southern Water in the UK, to explore and explain DataOps in more detail and understand how this role fits into D3. We are very grateful to Tariq for sharing this work with us.

Tariq starts with a simple initial scenario:

'At some point in your career there may have been a scenario when you are gathering requirements for some BI report, or Dashboard or anything that is data related and ask the question "when would you like this work to be done?" with the person answering "yesterday"
We are impatient. With the internet, we live in a generation where access to information is very easy.

Could you imagine if you were searching something on the internet using your favourite search engine you'll have to wait 24 hours before you get your results? Why shouldn't we expect the same for our analytics. With cloud computing and the ever-evolving technologies like machine learning and artificial intelligence, real-time data and processing has become more accessible and even more mainstream.'

Tariq Bhatti (2018)

Tariq proposes that DataOps is focused on six points, which you will see align with D3 thinking:

1. Culture

2. Governance and education (especially for data science)

3. Process and automation

4. Speed

5. Value

6. Focus on data

These six points he summarises, and therefore defines DataOps, as:

'Process-Oriented solution that would revolves around collaboration between operation (the business), governance and technology to bring data solutions that are integrated and automated for the business as well as delivering Value'

Tariq Bhatti (2018)

He goes on to provide some examples of DataOps in action and some of the things that might be delivered:

- *Creating synergy between business and technology by having data processes streamlined, improved and measured to break down data silos*

- *Empowering people to have the ability to move their idea from 'Proofs of value' to BAU in the shortest period of time*

- *To uplift companies to their next level their data journey. From prehistoric 'ETL Spreadsheets' to 'ETL SQL' or from BI to advance analytics like machine learning/artificial intelligence*

- *Showing analytics being a positive force that can transform an organisation*

- *Delivering analytics at a pace where it meets business needs on-demand*

- *Bringing data to the forefront to every company*

Figure 9.1 gives an overview of how DataOps works.

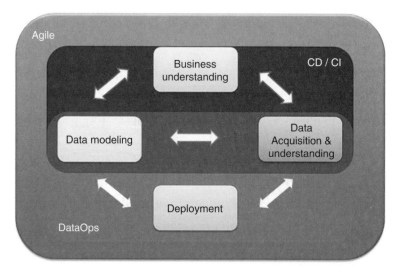

FIGURE 9.1 Overview of DataOps

'Business Understanding – This is a point where interaction between the business and the data function occurs. Understanding the use case, and business objectives, as well as thoughts and ideas on how to tackle a business problem or to deliver on business objectives

Data Acquisition and Understanding – A very important step, ensuring we know where we are sourcing the data as well as understanding it. As we all know, it's very hard to deliver value if we don't understand the data ourselves

Data Modelling – Having the business use case and the data, we can now start modelling the data, ensuring our model delivers in value

Deployment – Once the business has accepted the solution, the method of deployment can now be discussed. This would be a cross function solution between the business, the data team and IT

Using Agile to navigate between all areas of the process. Continuous Development/Continuous Improvement can help with the process between Business Understanding, data Modelling and data understanding analysis, while DevOps can help with Deployment'

Tariq Bhatti (2018)

The DataOps team requires, or must have access to, the following skill sets:

- Data architects

- Data engineers

- Data scientists

- Database administrators

- Data governance

- Data analysis

These skill sets may be deployed in a number of organisational structures, the one in Figure 9.2 was used at Southern Water in 2018 during the formation of the data team and is used here as an example:

Tariq concludes with a quick and easy way to differentiate Agile, CI/CD, DevOps and DataOps:

*'**Agile** focuses on **processes** highlighting **change** while accelerating delivery*
*CI/CD focuses on software-defined lifecycles highlighting **tools** that emphasise automation*
*DevOps focuses on **culture** highlighting **roles** that emphasise responsiveness*
*DataOps focuses on **data** highlighting **value** to the business as well as its lifecycle'*
Tariq Bhatti (2018)

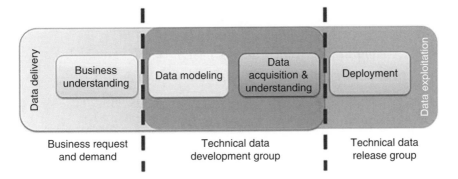

FIGURE 9.2 Example of DataOps skill sets

Technology

The final element of the triangle is 'technology'. Like the previous two elements, the technology in an organisation must be set up in such a way as to support dynamic data-driven transformation. It is possible to define and characterise this both in the positive and negative: what it is and what is shouldn't be.

For technology in an organisation to support D3 it should be flexible, responsive, forward looking and capable. It should have an ethos of 'interoperability', different pieces from different suppliers or different legacy being able to operate together. In many ways the technology should have an open architecture.

> '*Open architecture* is a type of
> computer *architecture* or software *architecture*
> intended to make adding, upgrading, and swapping
> components easy.'
> Wikipedia
> https://en.wikipedia.org/wiki/Open_architecture

'Vendor-independent, non-proprietary,
computer system or device design based on official
and/or popular standards. It allows all vendors (in
competition with one another) to create add-on
products that increase a system's (or device's)
flexibility, functionality, interoperatability, potential
use, and useful life.'
Business Dictionary
http://www.businessdictionary.com/

On the other hand, the 'negative' for technology can be summed up very easily; it should not be a barrier to transformation. It should be capable of supporting the ideas and imagination of the business and move at the pace the business wishes to move at. The alternative is unthinkable, and a reality that is rarely honestly addressed by an organisation. If the technology is not capable of effecting change as fast as the business requires and expects then, in these circumstances, an organisation is in a situation where 'change is constrained by technology' rather than being in a situation of 'data driven business transformation'. One is a recipe for failure, stagnation and being over taken by competitors, the other is innovative, market aggressive and winning.

Again, organisations may need to adopt a hybrid compromise, maintaining legacy technology to keep the existing business processes, and embracing new technologies and new technology approaches that genuinely support D3. The two should not conflict, they should complement and at some point in either the systems architecture or the data architecture they will almost inevitably converge. Achieving and supporting this hybrid approach may require new skill sets and new mind sets. So, an organisation may end up with two IT departments. We have noticed in our journeys and experiences that most organisations already have two functioning IT departments and two operating technology stacks; formal supported technology and shadow IT. The existing two, with shadow IT, are far more dangerous, risky and ungoverned for an organisation than a proposed new hybrid environment. So there are many reasons to adopt technology that supports D3, one of the most significant might be to do away with shadow IT. Shadow IT tends to occur when an organisation's technology is not capable of supporting the business needs and the pace of change that they wish to effect.

To some extent it has to be assumed that if the initial data-driven business transformation has started to move significantly to the first envisioned end state then the technology to support this must be in place to some extent. The battles with ageing, legacy technology should be over, already won or well and truly underway.

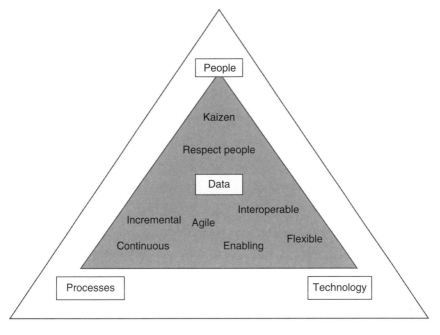

FIGURE 9.3 The dynamic data-driven business transformation triangle – D3

The Gartner report 'Predicts 2018: Analytics and BI Strategy' makes some interesting observations about 'technology':

*'Adaptability is at the heart of evolution, and survival.
In the case of technology, it must adapt to our
lifestyles, constraints and desires.'*
Gareth Herschel et al. (26 March 2018)

Technology must be responsive to the business needs, and not vice versa (see Figure 9.3).

Chapter 10

Dynamic Data-Driven Business Transformation – D3

A D3 approach encompasses changes to all three elements of the standard people, process and technology triangle, bringing them closer and making them more dependent on each other. It may be that a D3 methodology is rolled out or applied to one part of the business, and perhaps gradually spread out across the business over time once it is tried and trusted.

The incremental nature of D3 is very important – again, this is the start-up mentality. Start-ups rarely wait to achieve 100% product development before entering into a market; they release incremental versions, iterations. They learn from the market response, from the adoption and take up. They build KPIs to measure success and target improvements in these areas. A D3 approach is similar, it is dynamic, constant delivery is more important that viewing a fixed end state, so the transformation never ends. The growth of Amazon, Uber, Apple and many others fits into this approach of not striving for a fixed end state, they are continually growing, changing, bringing new products and services to market and interacting with their customers in new ways.

British cycling in recent years is another good example. Ian Drake, British cycling chief executive since 2009, had the view that cycling has been around as a competitive sport for a long time and that it was going to be for hundreds of years in the future. He was of the opinion that, even though he was only going to be looking after British cycling for a short time, he wanted to leave it in a better state at the end of his custodianship than he had found it in.

Drake's view is that we are obliged to make things better and move them forward, and that we should find a way to do this.

We have also seen the impact that Sir Dave Brailsford had on British cycling with his approach of small incremental improvements, that must be delivered as part of an ongoing and dynamic transformation.

When Sir Dave Brailsford became head of British cycling in 2002, the team had almost no record of success: only a single gold medal had been won in British cycling's 76-year history. That changed under his leadership. At the 2008 Beijing Olympics, the GB squad won seven out of ten gold medals available in track cycling, and they matched that achievement at the London Olympics four years later.

As a former professional cycler who holds an MBA, Brailsford applied a theory of marginal gains to cycling. He gambled that if the team broke down everything they could think of that goes into competing on a bike, and then improved each element by 1%, they would achieve a significant aggregated increase in performance.

Brailsford admits to applying Kaizen methodology to deliver transformation of the GB cycling squad. He aimed for small, not big, changes and adopted continuous improvement through marginal gains. It was very much an approach of not aiming for perfection, but rather focusing on progression and compounding the improvements. He identified the power of contagious enthusiasm. This sort of contagious 'leaning-in' can overcome some of the human barriers identified by Viki, and a great leader will build on this.

However, Sir Brailsford did recognise that the marginal gains approach doesn't work if only half the team buy in to it. He suggested that if only half the squad leans-in, the search for small improvements will cause resentment. However, if everyone is committed, there's mutual accountability, which is the basis of great teamwork. In other words, if there are cynics in the team then resentment will build up on both sides. The cynics will resent the success and the committed team members will resent the cynics sitting back and waiting for failure.

Now extend the same proven successful approach to include technology and processes in a business context and you are set up for embedded data-driven business transformation. The purpose of D3 is to 'unfetter' the data so that the business changing potential within it can be released.

Finally, to run the business in the new data-driven world an organisation should have a senior executive who is responsible and accountable for releasing the power held in the data. We have spent much of the last year on the speaking circuit, since the publication of our first book *The Chief Data Officer's Playbook*, answering the question 'why should an organisation need a CDO?' Our answer has been consistent and we often prove it with a show of hands in the auditorium:

> Every organisation accepts the need to have a CFO, to be responsible and accountable for the money within an organisation. Therefore, if data is an asset, and if data is going to enable or drive business transformation, then surely the organisation needs a senior specialist to be responsible and accountable for the data.

Perhaps the single biggest indicator that an organisation is prepared to run a data-driven business is the presence in their senior team of a CDO. Given the right D3 structures and approaches around her the business can deliver and sustain data-driven transformation.

I am sure most of us have seen the Hollywood movie that we mentioned briefly earlier, *Money Ball*, which demonstrates the significant influence that data played in transforming the Oakland As. If you've not seen it and are into data, watch it.

Let's consider another example. We may not all be golf fans, but the Ryder Cup team selections and tactics have been transformed by the use of data. The impact of data has been played out in recent Ryder Cup competitions, not least the most recent in 2018. It is interesting to think about and understand how the use of data might not only have impacted the result but also how the Ryder Cup teams were organised and run, how and where decisions were made. Sean Ingle (27 September 2018) in *The Guardian* described this very well when he wrote that Paul McGinley, the European team vice-captain in 2010 and 2012 was a statistician. Apparently, McGinley would arrive at meetings with bundles of paper, probably spread sheets, which contained data on how the players ranked on various golfing attributes. Ingle notes in an interview that McGinley says that the analytics in golf has come a long way since 2012.

Doesn't this remind us of organisations being run from the ubiquitous spread sheet? But clearly McGinley acknowledges that the use of data has evolved since 2012.

In the 2018 Ryder Cup it is was obvious that data was providing insights. The leadership of the European team were supported by six experts from 15th Club – while the American team were supported by the Scouts Consulting Group. After the US win at the 2016 Ryder Cup, team captain Davis Love recognised Scouts role as an integral part of the success.

It is very interesting to note that both the US and the European Ryder Cup teams decided to use external consultancies, as they didn't have the skills in-house. Perhaps this will change over time. In *Money Ball* the character played by Brad Pitt, the manager of the As, brought this skill in-house into the team.

So, to interpret this for the non-golfers; in 2016 the data guys told the US captain to set up that year's competition golf course, which was in the US, as an 'easy course, with wide fair ways' because this is what the data showed would suit the US players. Here we have the first piece of evidence of data-driven transformation in golf. But it wasn't only the US team that were using data.

McGinley is probably the first Ryder Cup captain to fully embrace analytics.

In 2018 the European team captain, Bjorn, was assisted by 15th Club in assessing the performance of each player and how best to use the insights gained to select the team for each day's play.

By 2018 the evidence shows that the European team had transformed. Hence, they were able to demonstrate dynamic responses to the data, and perhaps were the first to deploy a 'GolfOps' team driven by data.

Mark Broadie is often credited for the revolution in golf analytics. Broadie is a professor at Columbia Business School who analysed four million golf shots on the PGA tour between 2003 and 2012. In 2014 he published the book *Every Shot Counts* – and to quote from Amazon 'Columbia Business School professor Mark Broadie's paradigm-shifting approach that uses statistics and golf analytics to transform the game.'

So, before the 2018 Ryder Cup McGinley was asked by Ingle if data would make a difference to that year's competition. McGinley admitted that it was going to be close and he pointed out that even with statistics and data analytics the result would be down to the players.

No spoiler alert required, but the European team won the Ryder Cup in 2018. Perhaps McGinley was supporting our hypothesis about the importance of the people.

Isn't it time that we let data drive business transformation as well as golf transformation?

In summary, for an organisation to run in a data-driven world they have to understand and accept that there is no 'arriving at the destination'. The original 'vision' of the end state to be achieved by the data-driven business transformation is merely a light at the end of the tunnel that is never reached as aspirations grow and the opportunities change and new threats emerge. However, one of the deliverables of the original end state vision is the creation of a culture, an environment and a framework for continued dynamic transformation, perhaps better viewed as evolution. The essential levers in this dynamic framework are the people, processes, technology and, of course, the data.

Chapter 11

Conclusion

W e hope that throughout this book we have been able to share with you some of the passion we feel for taking care of our data properly and responsibly. Data-driven business transformation is a growing concept, and is increasingly being talked about and viewed as the next big thing or 'revolution'. What is interesting is that it isn't always the phrase 'data-driven business transformation' that is used, sometimes it is wrapped up in the story of the 4th industrial revolution, or it is incorrectly referred to as 'digital-driven transformation' and sometimes – perhaps causing the most misunderstanding – it is referred to as 'digital transformation'. This last term causes the greatest confusion because it seems to imply that all that matters, or is required, is a 'digital transformation'. It doesn't suggest or relate to any desired outcomes – such as business transformation. It implies a technology upgrade or modernisation as an ends in itself.

For us it always comes back to purpose, why are you trying to do something? Why are you trying to elicit a change? Data isn't there just for the sake of us all collecting more data, that in itself is not productive or useful. During the discussion in this book we have, we hope, made the case for how you can have data-driven transformations, projects or programmes but that the overall goal has to be to have a data-enabled organisation. A company that has governance and ethics built into how they treat their data can drive value at a phenomenal rate from that same data.

We have given you the tools to get you started but there are different levels that you can use them at.

The level that is most widely observed is the lowest level, the initial steps. At this level a business is 'transforming' a single business process enabled by data. This process may be a customer interaction, a manufacturing processes, a logistics process or part of regulatory regime. Whatever it is, this initial level addresses one business process in isolation. We would suggest that this is the level of data-driven business transformation approached by a relatively data-immature organisation.

However, it is the level of lowest business risk and may be used as a stepping stone, a lesson, a way of proving something or to try something more ambitious in the future. Don't worry if you are starting with this ambition – you are still ahead of the companies and organisations that aren't even beginning.

The next level up, approached by companies who are further through their data maturity, is a 'data-driven business transformation' that is intended to transform the whole organisation but reach an end state. A state at which point a 'vision' will have been reached or achieved and transformation will stop. A vision that has moved you forward and that you have delivered against is a great place to be.

The final level, which will only be attempted or contemplated by the most data-mature organisations, is the complete business transformation driven by data that moves into a state of continuous improvement, what we have called D3. A shift to this level of transformation will be seismic and fundamental but will set up an organisation to succeed and continue to succeed as markets, competitors, opportunities and challenges develop and continue to emerge both internally and externally.

Whichever level of transformation is attempted or delivered, the common denominator is 'data', that is, the insights in data and the use of data. The commonality of data ensures that the 'business', 'business outcomes' and 'business value' are the focus rather than the 'technology', and that the transformation is initiated and driven by the business using their data. Now, interestingly for a data-orientated book, we think it's important to remember that the data underpins and enables the business but that the people underpin the data and the transformation.

There are seven elements of a data-driven business transformation that should be reiterated and used as a constant guide to keep your data-driven transformation true.

First, the change should be dynamic and continuous, the 'end state' – the 'vision' – should constantly be evolving as capability and understanding grows, and as external forces exert pressures and reveal opportunities.

Second, a data-driven business transformation will require, and will deliver, a fundamental shift in culture. A change in culture not only towards data and how it is owned, used, managed and exploited, but also a shift in the culture of 'change' itself. This cultural change will support more agile methodologies to respond quickly to changing business need and to exploit opportunities quickly.

Third, a data-driven business transformation will need a particular form, style and leadership, although it may not always be the same style of leadership.

Fourth, there will be a move away from focusing on 'technology' and 'digital' with a growing focus on 'data' and the data within the technology and digital platforms. Perhaps what we are suggesting here is that there should be a rebalancing of the focus that has been stimulated by the growing understanding of the power and insight in the data.

Fifth, a data-driven business transformation should set itself up to succeed. The right skills will need to be recruited, and these are likely to change in nature as the transformation progresses.

Sixth, the organisation needs to be sure of its starting point. It must know its level of data maturity before the transformation begins to truly understand the challenges ahead and the course to be steered.

And, finally, remember about the people, they are your army for looking after, using and valuing the data. The people are the powerhouse of creativity that will mobilise the data, keep it safe and ensure that it is used ethically. We need to make sure we humanise the data transformation.

The steps we have taken you through – understanding your position, through to practical ways to make a change in each of the data elements, which in turn lead onto the ways of changing and embedding the differences – only work if you take the first step. Don't be afraid of data. We currently live in a culture where there is a fear of data and what it can lead to, whether that is misuses of data or the machines taking over. Become data cheerleaders like us and create a future where we leverage the real power of data. That power is the people we will save because clever machines and data help us humans to understand what the most promising fields of medicine are, our understanding of our universe being expanded because they help in mapping dark matter and more among the stars, the power in data is the difference we can make in our organisations and in our customers' lives.

Data is there to be used, don't ignore it. Understand the business need, look into the data for insight, understanding and opportunity to make the real, sustainable changes. Hold onto your sense of responsibility and unleash your imagination.

We believe that we live in exciting and challenging times. The world is changing, new companies are emerging and growing faster than has happened historically, while older more traditional companies are struggling to compete and survive. At the centre of this challenge and opportunity, whatever the prevailing political and economic environment or shifts, is data. And, in fact, in some cases, and in some ways, it is data that is shifting the political and economic environments. Organisations ignore the data at their peril, or they look to it for opportunity. One thing is for certain, the winners will be using data to drive change, transform and build the future.

INDEX

Note: Page references followed by *'fig'* refer to Figures